SERICULTURE AND RURAL DEVELOPMENT

SERICULTURE AND RURAL DEVELOPMENT

G. SANDHYA RANI
Department of Women's Studies
S. P. Mahila Visvavidyalayam,
Tirupati. (A.P.)

DISCOVERY PUBLISHING HOUSE
New Delhi-110 002.

First Published-1998
Reprinted-2011

ISBN 81-7141-419-2

Published by:
Discovery Publishing House
4831/24, Ansari Road, Prahlad Street,
Daryaganj, New Delhi-110002 (*INDIA*)
Phone: 3279245
Fax: 91-11-3253475

Printed at:
Sachin Printers, Delhi- 53

FOREWORD

Sericulture has the vast potential for income and employment genration in rural areas. In this way it is most suited for a labour surplus economy like India and more so for the small and marginal farmers. Realising this truth, several states of Indian union have been encouraging the development of sericulture. Further, there has been a tremendous growth of sericulture in Andhra Pradesh and Karnataka States.

The book'*Sericulture and Rural Development*' is of much policy interest, considering the potential contribution that sericulture holds for rural development in India. The book is well designed and is the result of very hard and organised work with clear presentation and well conceived method.

The book is divided into nine chapters. The first chapter emphasises the emerging importance of sericulture, lays down the objectives and spells out the methodology. The second chapter begins with the origin of silk and has a flavour of 'Newton seeing the Apple fall' when one comes across the passage as the Icocoon falling into the tea cup of Empress Si Ling Shi' in China in 2640 BC. It gives an over view of the employment and income generation capacity of sericulture. In the third chpater there is a brief description of sericulture in other countries which helps to understand the global position of India as a silk producer. The fourth chpater could be read as one of the very few spectacular success stories of state initiated planned sericulture promotion programmes which helped to achieve significance production of raw silk in India during the period 1981-94. The success of sericulture in Andhra Pradesh and Karnataka states and prospects of sericulture particularly for small and marginal farmers is presented in chapter five. The presentation of 'income shares' gives an impression that the Cocoon producer ist the biggest gainer although the gains may be disproportionately in favour of traders. The growth rates presented in the chapter have effectively brought out the trends in the progress of sericulture. Seventh Chapter highlighted employment generation through sericulture which is both impressive and realistic. Eighth chapter assesses the problems like diseases and pests, fluctuations in prices etc., all of which suggest some of the teething problems of any fast growing sectors.

The book is a clear contribution to the understanding of the place of sericulure in not only the rural development of India but also in export promotion.

The present work offers vision, perspective and new dimension in rural economics as sericulture is an excellent and rewarding agro-based activity to the farmers in dry regions in India.

N. NARAYANA
Director
Centre for Rayalaseema Development Studies
Sri Krishnadevaraya University
ANANTAPUR (A.P.)

PREFACE

Countries like India have still to rely largely on the agriculture sector for economic development. Since agriculture has been regarded as a pillar of the national economy it has occupied a vital place in the development strategies of India. However in recent years Indian agriculture is on the threshold of a stage of development characterised by a shift from static technology to modern technology.

In this context, sericulture with its vast potential for income and employment generation in rural areas plays an important role in alleviating rural poverty. It is one of the crop enterprises identified as the most appropriate labour intensive cottage industries. This activity combines both agriculture and industry. Sericulture is best suited to India, where manpower and land resources are in surplus. By creating more employment opportunities in rural, urban and semi-urban areas it not only arrests rural migration but also promotes a series of cottage and small scale industries. 'Silk' the final product of sericulture enterprise, has very good association with the customs and traditions of the people living all over the world. Further, with improvement in the economic conditions of the people, the demand for silk is also increasing within the country and abroad.

Historically, China is the motherland of silk. Chinese treasured the secret of silk for centuries together, but gradually it spread to Japan, Korea, India and other parts of the world. Sericulture enjoys rapid growth mostly in South-East Asian countries and nearly 87 per cent of the world's raw silk production is from these countries only.

The history of sericulture in India is as old as the history of Indian civilization. In its long journey towards achieving prosperity it has experienced many ups and downs. However, during the last 20 years, India has made tremendous progress in the production of silk for which, there is an increasing demand. As it combines both agriculture and industry sericulture has been occupying a very important place in transforming the age old Indian agriculture into a modernised agriculture by intensive use of land and capital.

Sericulture in India is broadly classified into two district sectors namely mulberry and non-mulberry. Though, India has the unique distinction of producing all the four commercially known varieties of silk, over 90 per of the silk produced in India is mulberry silk. In recent years, sericulture has achieved enormous progress in evolving suitable mulberry varieties and techniques to bring about new silkworm races. As a result, the productivity has increased and sericulture has become a highly remunerative activity. Further, this industry which was confined to only five States namely Karnataka, Andhra Pradesh, West Bengal, Tamilnadu and Jammu and Kashmir has spread to almost all the states of India. During the recent past, of all the states Karnataka has emerged as the leading producer of silk, producing more than fifty per cent of the mulberry silk production in the country. This state is now regarded as the 'Silk Bowl of India'. Next to Karnataka, Andhra Pradesh though not a traditional state in silk production, occupies the second place in the country in producing mulberry raw silk.

In Karnataka, which ranks first in raw silk production, the area under mulberry cultivation increased from 116.19 to 160.84 thousand hectares, reeling cocoon production increased from 38.01 to 70.21 thousand tonnes and raw silk production from 2.88 to 8.25 thousand tonnes during 1980-81 to 1993-94. At the same time in Andhra Pradesh area under mulberry cultivation has increased from 18.90 to 90.77 thousand hectares, reeling cocoons production increased from 10.98 to 24.51 thousand tonnes and the raw silk production from 0.08 to 2.86 thousand tonnes.

The National Sericulture project (NSP) and many other projects and programmes which are under implementation in these two states are also responsible for the commendable growth of sericulture. The promotional agencies have been playing a crucial role in the development of sericulture in India. The central silk Board, Bangalore and the Central Sericultural Research and Training Institute, Mysore are helping it by providing suitable technologies, evolving new varieties of mulberry and silkworm races, taking research findings to the farms and giving training to the sericulturists.

In Andhra Pradesh, sericulture industry is mainly concentrated in the Rayalaseema region. Even in Rayalaseema, Anantapur is popularly noted for sericulture and occupies first place in the state. Similarly in Karnataka, sericulture has struck deep roots in Kolar district. In Kolar, sericulture is regarded as a traditional occupation and closely interwoven with the life styles of people of this district. These two districts have a unique distribution of similar agro-climatic conditions because of their geographic situations. Mulberry cultivation in these two districts is being undertaken under irrigated conditions. Mulberry, being drought resistant, recurring droughts in these two districts have not prevented the tremendous growth of sericulture. Thus they have become pioneers in sericulture in their respective states. Sericulture in Anantapur has gained momentum with the implementation of the DPAP.

In undertaking this study, an attempt has been made to analyse and review the growth and development of sericulture in Andhra Pradesh and Karnataka with special reference to Anantapur and Kolar districts respectively. An attempt has also been made to give a comprehensive idea of sericulture in rural economy. It is also intended to analyse the existing production and marketing practices and problems of sericulture in Anantapur and Kolar districts and to examine the cost and net returns of sericulture. A comparative study of the employment opportunities generated from sericulture and other principal crops in the study area reveals that the generation of mandays on one acre of mulberry garden is more than that of paddy, groundnut and ragi. This analysis clearly establishes the importance of sericulture over other crops in the generation of fresh employment opportunities in rural and urban areas.

A growing enterprise like sericulture, is also not free from certain problems like diseases which infect silkworms, non-availability of layings, climatic hazards, wide fluctuations in cocoon prices and also inadequacy of extension services to certain extent.

The study is confined to both Anantapur and Kolar districts and based on both primary and secondary data. The data, however, is drawn mostly from the field investigations carried out by the Researcher. The study has highlighted the prospects of sericulture in the districts and brought to light the inadequacies in production and marketing of the produce of sericulturists (i.e. reeling cocoons). It has also provided basis for offering specific suggestions for all round development of sericulture in India.

ACKNOWLEDGEMENTS

In bringing aout the book to a successful completion, I have received generous help from a number of people. I place on record here my indebtedness to all of them.

With immense pleasure and profound sense of gratitude, I take this opportunity to express my heartful thanks Dr. N. Narayana Professor of Economics, Sri Krishnadevaraya University, Anantapur for his constant encouragement and timely suggestions in carrying out this work.

I deem it my duty as well as pleasure to express my sincere thanks to Dr. H.G. Hanumappa, Professor of Economics, ISEC, Bangalore, for his valuable suggestions. I am beholden to Dr. Balakrishnama Naidu, Lecturer, Department of Econometrics, S.V. University, Tirupati for his help in adoption of relevant statistical tools.

My sincere thanks are due to the Commissioner of Sericulture, Government of Andhra Pradesh, Hyderabad and also to the Director of Sericulture, Government of Karnataka, Bangalore, for providing me the necessary data. My thanks are due to Sri Venkateswarulu, Additional Director, Sri Prabhakar Joint Director, Department of Sericulture, Hyderabad and also to Sri. Kanyadi, Joint Director of Sericulture, Bangalore, for their constant encouragement throughout this endeavour.

No words are sufficient to express my heartful thanks to my husband Sri P. Subramayam, Assistant Director, Ground Water Department. Government of Andhra Pradesh, for his encouragement throughout this work. He identified himself totally with the investigation and stood as a bulwark of support through all its phases.

I am grateful to the authorities of Sri Krishnadevaraya University, Anantapur for permitting me to publish my doctoral thesis.

I also express my sincere thanks to the authorities of Sri Padmavathi Mahila Visvavidyalayam, Tirupati. and to all my colleagues in the Department of Women's studies, for their encouragement and co-operation in this endeavour.

G. SANDHYA RANI

CONTENTS

CHAPTER 1
Introduction

In a majority of the developing nations of the world developmental efforts in the last decade have not been as successful as expected. Except in a few newly industrialized countries, in all the other developing nations poverty has been on the rise, economic growth has slowed down, employment faltered, and inflation is on an upward swing. Consequently the general levels of living are low for the majority of people in these economies. This is true not only in relation to their counterparts among the rich nations, but often also in relation to small elite groups within their own societies. These low levels of living are manifested in the form of low incomes, inadequate housing, poor health, low productivity etc.

A greater portion of the population of developing nations depends on agriculture, the primary sector, for livelihood for the simple reason that the first priority is for food and clothing. But the productivity levels in the agriculture sector, are low. This is not only because of pressure on land but agriculture in these nations is also often characterised by primitive technologies, poor organisation and limited capital and investment. Hence the developing countries are directing all their development strategies on raising the productivity levels in rural areas in order to achieve higher economic growth and better distribution of income. The majority of the rural agricultural farmers are small and marginal. Moreover, it is realised that these small and marginal farmers constitute a resource base for major agricultural programmes to increase the productivity and incomes of the farmers, thereby provide significant boost to the national economic growth as well as improved income distribution within the country.

The developing countries have to focus all their development strategy on raising the agricultural productivity. Countries like China and India have already shown that agricultural practices should be integrated with animalhusbandry, dairying, fisheries, horticulture and sericulture to generate more income for each household.

In India nearly 76.3 per cent of the population is living in rural areas and 68.8 per cent of it constitutes agricultural population.[1] Agriculture is a biological industry and is of fundamental importance in determining the economic development of a country. Agriculture in India has been a pillar of the national economy.

In India the particular importance of agricultural development came to be recognised as vital during the Post-independence period. Although great strides have been made in agriculture since independence, the desired level of production has not yet been achieved. The annual growth of agricultural output was far low, at 3.2 per cent when compared to 6.3 per cent in the industrial sector and 6.7 per cent in the service sector during 1980-91.[2]

Agriculture being the traditional occupation in India for many generations, an attempt is made here to identify the broad stages in the evolution of agriculture in this country from the primitive stage to the present modernised farming. These stages can be classified under three headings. The first is the primitive stage, mostly meant for subsistence and is also known as 'peasant' farming. For a long time, India has remained in the stage of peasant farming which is traditional in nature and most of the produce is intended for the peasant's own consumption. Output and productivity levels are low since simple and traditional methods and tools of farming are used. Capital investment is minimal while land and labour are the principal factors of production. The law of diminishing returns is in operation as more labour is applied to limited land. Labour is underemployed for most of the year, although they are fully occupied during seasonal peak periods such as planting and harvesting. The peasant usually cultivates as much as his family could manage without the need for hired labour. This type of subsistence agriculture is a highly risky and uncertain venture. In regions where farms are extremely small and cultivation is dependent on highly variable rainfall the average output will be low in poor rainfall years and the peasant and his family will be exposed to the danger of starvation. In such circumstances, the main motivating force in the farmer's life may be maximization of his family's survival chances rather than its income.

The second stage of evolution of agriculture might be called 'diversified or mixed' farming, where part of the produce is grown for self-consumption and part for sale to the commercial sector. Diversified or mixed farming represents a logical intermediate step in the transition from subsistence to specialised production. In this stage the staple crop no longer dominates the farm output since new cash crops such as fruits, vegetables, tobacco, coffee and tea are established together along with simple animalhusbandry. Staple crops occupy the land during some parts of the year and new cash crops can be introduced in the slack season to take advantage of both idle land and family labour. The farmer can thus have a marketable surplus which can be sold to raise his family's standard of living. Diversified farming can also minimize the impact of staple crop failure and assure a secure minimum income to the farmers.

The third stage of agriculture is represented by 'modern' farming, in which it is exclusively engaged in high productivity and specialized agriculture geared to the commercial market. It is the most prevalent type of farming in advanced industrial nations. On specialized farms the growing of food chiefly for the family with some marketable surplus is no longer the basic goal. Instead, it is purely commercial agriculture and profit becomes the criterion of success. Maximum yields per hectare are derived from the exploitation of natural resources as well as man-made devices such as irrigation, fertilizers, pesticides, hybrid seeds etc. The entire production is virtually for the market. For all practical purposes, specialised farming is in no way different in concept or operation, from a large industrial enterprise.

Although one can see all these three types of farming practices coexisting in India, peasant subsistence farming and small-scale mixed farming still dominate. Transition from these farming systems to commercial enterprise, farming is difficult in India because of a very large number of small-size operational holdings and lack of capital. Therefore, the improvement of small and medium scale mixed-farming practices will not only raise farm incomes and average yields but also effectively absorb underutilised rural labour and open avenues for achieving real and people-oriented rural development.

Indian agriculture, is mainly classified into two categories on the basis of availability of water resources, namely (i) rainfed or dryland farming and (2) irrigated farming. Rainfed agriculture depends on the monsoons. Indian agriculture is primarily rain dependent, and rainfed agriculture supports 40 per cent of the total cultivated area. India has a net sown area of over 140 million hectares of which only 41.2 million hectares are under irrigation. Irrigation facilities are inadequate in India. Therefore, agriculture is still a gamble and at the mercy of the monsoons. Generally speaking, 762 mm of evenly distributed rainfall is most favourable for normal agricultural growth. But, rainfall has been erratic in India. Though huge investments have been made to increase the irrigation potential, in the contrary, only 42 per cent of the total cultivated area is being cultivated as the facilities are far from adequate.

Moreover, even for most of our irrigation projects, rainfall is the only source. Owing to erratic and inadequate rainfall most of the dams, tanks and other water resources go dry during the summer season. The rainfed areas are frequently afflicted by periodic droughts, soil erosion, crop fluctuation and other related problems. To overcome drought problems and to improve the status of farmers, the Government of India has given high priority to major irrigation projects in the Five Year Plans. Though, the irrigation potential as a result has improved to a certain extent, the major irrigation projects have played only a limited role in the development of the agriculture sector. So with a view to improve the drought prone areas, the Government of India has devised and implemented such projects as are helpful for economic development, rural upliftment, eradication of poverty and unemployment.

The incidence of poverty and unemployment in rural areas is very high with 33.4 per cent of the people living below the poverty line, compared to 20.1 per cent of their urban counterparts.[3] The conditions of deprivation in which they are caught, no set of statistics can adequately describe. The truth is that poverty is an intolerable assault on human dignity and decency. Malnutrition, diseases, illhealth, illiteracy and unemployment pervade the lives of these vast population.

Since, a majority of the rural agricultural farmers are small and marginal, maldistribution of wealth, assets and unequal distribution of natural resources have led to regional and intra-regional disparities. Hence, the number of landless labourers, small and marginal farmers falling below the poverty line has increased, particularly in regions which have a slow rate of growth in agriculture. The land-holdings of small and marginal farmers which are below two hectares each and often fragmented into small pieces, pose many problems. The percentage of cultivable land per head has also decreased from 0.48 hectare in 1981 to 0.25 hectare in 1991 due to rise in population.

As per the 1991 census, the population of India is about 845 million, and India constitutes the second largest populous country in the world.[4] Indian population is expected to touch 986 million by 2001, of which rural and urban population are expected to be about 660 million and 326 million respectively. The labour force available in the age group of 15 to 59 is expected to increase from about 270 million to 380 million by 2001. The explosively increasing population has been putting tremendous pressure on agriculture with what one may term as diminishing returns over the years. Further, it is intensifying the problem of unemployment, underemployment and disguised unemployment. Infact, the problems of unemployment and underemployment are already severe and sure to become worse as the rate of growth of the labour force accelerates fast. The problem of unemployment and underemployment is acute in rural areas and those who suffer most from it are agricultural labourers, rural artisans and small and marginal cultivators.

Suitable strategies are needed to overcome the above mentioned problems. In India from the beginning, the Five Year Plans have been stressing the need for enlargement of work opportunities for the growing millions. This has been one of the supreme tasks of planning. Typical is the statement in the draft outline of India's First Five Year Plan that a 'development plan' is essentially an effort to create conditions for full employment.[5] Despite the emphasis on creation of employment opportunities with the more sophisticated employment targets fixed, unfortunately the results achieved fall very much short of the announced targets and goals. Inspite of massive investment undertaken since the wake of planning, over the many years a large proportion of India's population continues to live under grim conditions of poverty, and unemployment.

Infact the actual application of the better methods of agriculture may depend very much on such structural measures as land reforms, better systems for the provision of credit, better land use, conservation of basic resources like soil and water and formulating contingency management plans to meet seasonal aberrations. Schemes and programmes relating to agriculture adopted in the past were either intended for meeting shortages or for political gains. Hence, what is needed is the development of agriculture for the full realisation of its potential so that the quality of life of the rural masses may be improved.

Indian policy-makers and planners have worked hard and searched for effective measures to abolish unemployment and poverty. Moreover they have realised that the employment opportunities created must be productive enough in ensuring minimum income and standard of living to the poor. As the capital intensity of urban industry is greater than that of rural occupations, rapidly increasing urbanisation and industrialization cannot guarantee adequate employment opportunities to the growing labour force. Hence, the solution lies in accelerating rural development and generating productive employment, despite the trend of increasing pressure of population on land.

Therefore, the discovery of productive employment opportunities under integrated rural development assumes vital importance for the economic development of India. In this context it is quite natural to start with agriculture as the biggest of India's industries. The most important method of securing an increase in agricultural output is to induce the cultivators to adopt better agricultural practices.

For creating more employment opportunities to absorb the surplus labour in the agricultural sector, land augmenting technology is the best alternative strategy. As the supply of land is purely inelastic, employment generation in agriculture could be augmented by mainly increasing the intensity of labour to land. If all the factors of production are properly co-ordinated, then the expansion of employment opportunities would naturally increase. But in India even in irrigated areas the per hectare labour output is low. Thus, there is a strong complementarity between growth of agriculture output and employment generation in Indian agriculture.

In fact, Indian agriculture now is no longer confined to the cultivation of traditional crops alone. Farmers are encouraged to take up activities in non-farm sectors such as animal husbandry, poultry, fisheries, social forestry, sericulture etc. Hence, the planning for employment in agriculture sector should aim at tapping the potentialities in these allied activities.[6]

It should be noted that employment and income are the two faces of the same coin. The income of the family could be raised by (1) increasing the labour force in the family itself, (2) eradication of unemployment, (3) raising the daily earnings and (4) involving the family members in income generating activities. The first measure may be ruled out because, as far as the poor are concerned it is beyond doubt that their labour participation rate as well as the labour force days per person are as high as humanly possible.[7]

Therefore, it is better to create more employment opportunities, where there is scope for doing so. As rural areas are rich in natural resources the large employment opportunities in these areas could be created by (1) increasing the area under cultivation, (2) increasing irrigation facilities, (3) encouraging intensive cultivations and multiple cropping, (4) increasing the use of chemical fertilizers and HYV seeds and (5) introducing modern technology and co-operative farming for the benefit of small and marginal farmers and finally (6) searching for alternative crops which are labour intensive and yield more income. As there is no integration between agriculture and agricultural processing activities, the plan designed to aim at the augmentation of employment and income opportunities in rural areas should have a micro-level base enabling planning of various activities in a mutual supportive manner.[8] There is a view that the promotion of employment through the use of labour intensive technology would increase the per unit employment and may result in reduction of profit. But this need not be so in the case of biological industries where there is no competition from machine-made urban based products.[9]

Hence, changes in employment and income opportunities in rural areas may be brought about by selecting a highly labour intensive and high income yielding cropping pattern suitable for both wet and dry cultivation. Change in the cropping pattern is decided on the basis of soil fertility, climatic factors, rainfall and irrigation. Adequate irrigation facility is a prerequisite for the development of agriculture, as it makes for higher level of employment and income generation. It should be noted that irrigation and, even more importantly, shift in the cropping pattern towards a more labour-intensive and income-yielding crop cycle are responsible for high level of labour inputs.[10]

IMPORTANCE OF THE STUDY

It can be said unhesitantly that Indian agriculture is on the threshold of a stage of development characterised by a shift from static technology to a modern technology, in which capital requirement and purchased inputs occupy a large share. Much of the success of the new programmes will depend upon the ability of the workers to act as growth promoters. So all the new technologies should be built on the ability of the farmers to understand and adopt the new agricultural activities which will ensure higher income and employment to the rural population.

It is in this context, sericulture, with its vast potential for employment generation in rural areas, plays a vital role in reducing rural poverty and unemployment. It is one of the crop enterprises which are identified as most appropriate labour intensive household activities. It combines both agriculture and industry. It provides gainful employment not only at the stage of the production of mulberry leaves but also at the stage of the rearing of silk worms using the output of the former as an input of the latter. Sericulture has been playing a very important role in transforming the tradition bound agriculture into a modernised agriculture by intensive use of land and capital. Employment potentialities of different crop enterprises contributing to the growth of agricultural income can be estimated by working out the ratio of land use by a particular crop enterprise and the employment generated within the crop enterprise as well as in the processing stage of that crop in the same production line. Sericulture in which production of mulberry at the cultivation level is integrated with the rearing of silk worms by the same farm households, provides scope for augmenting employment opportunities and also increasing the income levels of farm households.

Now, in India sericulture has become the most promising rural activity due to various reasons. It generates direct and indirect employment. First mulberry cultivation creates employment on the farm and secondly cocoon production which uses the mulberry leaf as an input creates largescale employment for the family labour of the mulberry growers if that operation is also undertaken by the same household to reduce its underemployment in agriculture. There are, however, instances of non-mulberry growers taking up cocoon production alone as a full-time occupation. They buy leaves from the mulberry growers and use them as raw material for cocoon production. Reeling is also undertaken mainly in rural or semi-urban areas and the employment thus generated would help to reduce significantly rural unemployment. In short, sericulture as a whole, by the very nature of its activities, creates large-scale employment and income generation opportunities in rural and semi-urban areas close to each other accelerating their economic growth.

NEED FOR THE STUDY

The history of sericulture in India is as old as the history of Indian civilization. In its long history it has experienced many ups and downs. However, during the last 20 years, India has made tremendous progress in the production of mulberry silk for which there is an increasing international demand. There is tremendous scope for the expansion of its production in the country.

In recent years sericulture has achieved enormous progress in evolving suitable mulberry varieties, and techniques to bring about new silkworm races suitable for tropical climatic conditions. With the evolution and introduction of more productive silkworm races, the productivity has increased and sericulture has become a highly remunerative activity. Attracted by these advantages more and more farmers have taken up sericulture and the industry has spread to almost all states in India. However, Karnataka has been the leading producer of mulberry silk accounting for more than 50 per cent of its production in the country. This state, is now regarded as the 'Silk Bowl of India'.

Mulberry cultivation in Karnataka is concentrated in the districts of Mysore, Bangalore, Kolar, Tumkur, and Mandya. Mulberry in the Mysore district is grown mainly under rainfed condition whereas in the Kolar, Bangalore and Mandya districts it is mainly grown under irrigated conditions. Originally, the mulberry crop was introduced in the Mysore district and its surrounding areas as a dry-crop. Subsequently it was grown as an irrigated crop in certain concentrated pockets of Kolar and Bangalore districts. The entry of Mandya district into the field is recent. Presently mulberry has begun to spread to certain restricted pockets of the Tumkur district also. This is partly due to the efforts made by the Departmental of Sericulture in Karnataka to propagate mulberry crop and cocoon production in the non-traditional areas and partly due to the trial-and-error efforts made by the farmers in those regions who were attracted by very high profit margins in cocoon production. In 1951-52 the area under mulberry cultivation was only 43.2 thousand hectares, which increased to 1.60 lakh hectares in 1993-94. Accordingly the production of cocoons also recorded a rise from 1182 metric tonnes in 1951-52 to 70,210 metric tonnes in 1993-94. The quantity of silk produced in the state also recorded a remarkable increase from 727 metric tonnes in 1951-52 to 8,250 metric tonnes in 1993-94.[11] This shows that sericulture in Karnataka is growing at a very high rate. The farmers in the non-traditional areas of the state are quick in seizing the opportunities before them and have accepted mulberry cultivation and silkworm rearing as a highly profitable enterprise.

Andhra Pradesh comes next to Karnataka in producing mulberry raw silk, though it is not a traditional silk producing state. In Andhra Pradesh, sericulture has proved to be a money spinner for many middle-class families. Except the Ranga Reddy district, almost all other districts in the state have taken to sericulture activity. In 1956, mulberry cultivation in the state was only in 1,212 hectares and now it went up to 90,800 hectares by the end of 1993-94. Once confined to a few pockets in the Anantapur and Chittoor districts bordering Karnataka, sericulture has caught up in such a big way, that in just 10 years time the state of Andhra Pradesh has come to rank next only to Karnataka, the premier state of silk. About five-lakh farmers of all categories, big, medium, small and marginal, are engaged in raising the mulberry crop making use of the available subsidies and liberal loans from financial institutions. The cocoon production in the state went up to 24,510 tonnes by the end of 1994.[12] Though, almost all districts in the state have taken up sericulture, the main concentration of mulberry silk production is in the four Rayalaseema districts namely Anantapur,

Cuddapah, Kurnool and Chittoor. Of them Anantapur stands first in the entire state in producing mulberry raw silk, and Chittoor comes next.

At present, Karnataka and Andhra Pradesh show enormous growth rates in sericulture development. Andhra Pradesh has been ambitious of achieving the first rank in the production of mulberry raw silk in India. If the present tempo of growth is maintained, it would not be a surprise if it achieves its desired aim before long. Therefore, it is both necessary and worthwhile to study sericulture development in these two leading states and what further prospects it has. The present study focusses its attention on identifying the crucial aspects of sericulture like mulberry cultivation, cocoon production (which are entirely rural based activities) and employment generation through these two activities.

REVIEW OF LITERATURE

The tremendous growth of sericulture in Andhra Pradesh and Karnataka states has led to many research studies and surveys by various institutions. Some of these studies are general in nature and some are scientific. In the following paragraphs an attempt is made to give a brief account of the studies made so far on various aspects of sericulture in India.

There are a number of book-length studies of sericulture. D.L. Narayana's book, 'Economics of Sericulture in Rayalaseema' (1979) is an elaborate study of sericulture in the four Rayalaseema districts, namely Kurnool, Cuddapah, Chittoor and Anantapur of Andhra Pradesh. This study points out that sericulture is mainly concentrated in the Chittoor and Anantapur districts. It gives a detailed account of the economics of sericulture, the problems of sericulture and also the scope for its development.

D.V. Ramana's study 'Economics of Silk Industry in India' (1987) gives a picture of the sericulture activity during the eightees, and covers both its agricultural side and industrial aspects. The author discusses the role of sericulture in economic development and the status of sericulture and silk industry in the world. He also presents an analysis of the economics of sericulture bringing out its importance in terms of income and employment generation.

S.B. Shantha Raj Kumar in his work 'Silk Handloom Industry in Andhra Pradesh', (1986) examines the pattern of work and conditions of weavers, the process of manufacturing the warp and weft etc. He also deals with the organisation of silk business from raw silk to the final product. He also offers some suggestions to better the lot of weavers.

Abdul Aziz and H.G. Hanumappa, in the book, 'Silk Industry, problems and prospects', attempt an overall study of the silk industry. It is a collection of essays by different scholars. In this book various operations and activities relating to sericulture and its development are discussed.

Sanjay Sinha's book, 'The Development of Indian Silk', gives a clear picture of sericulture in the Indian economy. The position of sericulture in India, the products and production systems, the policy and micro-economic issues in silk production are discussed in this book.

A.R. Raj Sinha and K.V. Govinda Raj, in their book titled 'Employment and Income in Sericulture', present sericulture as a tool which helps to increase employment opportunities in rural economy. The role of sericulture in generating employment and income is discussed in a very clearcut manner. The creation of employment and income in silk reeling units is also dealt with.

T.D. Koshy, in his book, 'Silk Exports and Development' (1993), has tried to educate the silk exporters about the products they deal with. Part - I of this book deals with various aspects of sericulture like silk production, processing and procurement and part - II is about silk trade and export procedures.

'Sericulture for Rural Development' (1986) edited by Hanumappa comprises eleven papers presented by eminent professors and scholars, which highlight sericulture in Karnataka right from mulberry cultivation down to research development and training activities. The book as a whole gives a clear picture of the role of sericulture in rural development.

'Culture and Sericulture' (1982) by S.R. Charssley dwells on the significance of the livestock industry and sericulture, based on agriculture which in India have been subject to schemes of intensive development.

Mahesh Nanavathy, in his book 'Silk Production, Processing and Marketing' (1990) gives a clear picture of silk producing activities and a detailed scenario of the history of silk in the world.

P. Venkata Narasaiah in his book 'Sericulture in India' (1992) examines in detail the growth of sericulture in India. He points out that sericulture, which is an agro-based cottage industry, fits very well in India's rural structure, where agriculture continues to be the main industry. This book contributes significantly to a better understanding of the strengths and weaknesses of sericulture operations. It also offers some useful practical suggestions to overcome the various current problems and hindrances faced by the industry, and to ensure overall development.

The book 'Sericulture Society and Economy', (1993), edited by H.G. Hanumappa, is a volume of ten research papers by scholars engaged in studying the importance of sericulture to our economy and society. H.G. Hanumappa, who is the principal contributor to the volume, presents sericulture as an economically rewarding enterprise. He says that sericulture, consisting of several sets of activities, offers immense scope for social scientists to explore the interface that sericulture has with economy and society. These essays offer an insight into the processes by which rural development is taking place. The book is divided into two sections. The first dealing with the impact of sericulture on output, income, energy and rural institutions. The second section consists of papers examining the socio-economic issues of the silk-reeling aspect of sericulture.

Dr. Mrs. G. Ganga and Dr. Mrs. J. Sulochana Chetty's book 'An introduction to Sericulture' (1991), gives an overall idea of sericulture. In this book the authors discuss in detail the history of sericulture, the importance of sericulture and present a package of practices for mulberry cultivation. They also give an account of the diseases and pests generally to which silk worms are prone. The physiological aspects of silk worms were also given in detail.

The book 'Sericulture and Development' (1993) edited by J. Acharya gives a succinct report of his field studies in sericulture in the southern states of Karnataka, Andhra Pradesh and Tamil Nadu. The essays cover the whole gamut of issues concerning sericulture and development, women and children, technology and extension pattern of sericulture in different agro-climatic zones, problems of sericulture under water scarcity etc.

In addition to the above books, there are a number of papers by eminent scholars expressing their views favouring sericulture as an effective tool for the eradication of poverty, unemployment and for raising the standard of living of the rural masses.

In the paper on 'Issues in Sericulture activities - Macro perspectives", Prof. H.G. Hanumappa and Mangala, (Institute for Social and Economic

Change, Bangalore,) present brief note on the nature and extent of sericulture and allied activities in Karnataka. They also point out that sericulture has served as the main as well as a subsidiary occupation to thousands of agriculturists in Karnataka. Thereby this state has come to occupy a prominent place in the sericultural map of India having 75 per cent of the mulberry land and 78 per cent of the total silkworm rearing and 68 per cent of raw silk reeling in the country.

In his paper, 'Concurrent evaluation of larger Rural Development Projects - Some Experiences with Karnataka Sericulture Project', by H.G. Hanumappa, expresses his views on sericulture as a tool for the upliftment of rural poor. He suggests that employment opportunities in rural areas would be increased by changing the cropping pattern from less labour intensive crops to more labour intensive crops leading to high level incomes.

L. Devasurappa in his paper, 'Silk Industry in Karnataka', gives a general idea of the performance of the silk industry in Karnataka. He presents a picture of the origin and growth of sericulture in the State. The various activities involved in sericulture commencing from mulberry cultivation to silk weaving are discussed in detail.

K. Veeraiah in his paper 'A critical look at the mulberry cultivation in Karnataka', gives a brief account of mulberry cultivation and cocoon production. The important factors that have contributed to the stability of the cocoon crop are discussed in this paper.

'Mulberry Cultivation' by Boraiah covers not only mulberry silk but also Eri, Tasar and Muga silks. The authors however, gives top priority to mulberry cultivation and considers some ways of successful cultivation of mulberry gardens to obtain high productivity.

H.G. Hanumappa, in his paper, 'Economics of sericulture - Micro-perspectives', studies in detail the parameters of economic viability of sericulture. According to him the extent of economic viability accomplished in a crop enterprise like that of mulberry cultivation followed by silkworm rearing depends on (i) increasing the productivity of land by the adoption of improved mulberry varieties and (ii) improved practices in silkworm rearing by opting for better and disease-free layings, maintaining hygienic conditions for rearing, etc.

T.N. Sonwalkar, Director, Central Silk Technological Research Institute, Bangalore, in his paper "Factors Influencing Reeling, Efficiency", gives some reasons for the low productivity of cocoons. He also suggested some measures to overcome this problem.

There are some papers which deal with the purely biological aspects of sericulture. M.N. Narasimhanna, Director, National Silkworm Seed Project, Central Silk Board, Bangalore has two papers. In one of them, 'A Sound Seed Organisation vital for Sericulture Industry' he discusses the organisation of seed in India and advanced countries like Japan, the then U.S.S.R. and China. He emphasises that a three-tier seed multiplication programme, systematic moth examination and supply of young age silkworms alone can bring the country better prosperity through sericulture.

In the other paper, titled 'Silkworm Seed Production', Narasimhanna explains how silkworm seed production can be managed efficiently in a tropical country like India. Poor quality of silkworm seed reduces the cocoon yield and silk recovery. As a result the net return will be poor and sericulture cannot compete successfully with other cash crops. Hence, efficient management of seed production is needed.

There are some studies on the problems and prospects of sericulture development. G. Thimmaiah and V.M. Rao in their paper 'Problems and Prospects of Sericulture Development in Karnataka - A field review', discuss various problems faced by the sericulturists in Karnataka. They also present a sketch of the prospects of sericulture development in the state.

In the paper entitled 'Problems and Prospects of Sericulture', K. Periswamy discusses many problems of sericulture, commencing from mulberry cultivation to cocoon production. He suggests some techniques and measures to overcome these problems. He also gives an account of the status of sericulture in China, the leading country in mulberry silk production.

The Central Silk Board, Bangalore, the Central Sericultural Research and Training Institute, Mysore and National Institute of Rural Development, Hyderabad, have conducted many workshops, seminars and surveys on sericulture development and published their reports. These reports are very useful in formulating an idea of the growth potential of sericulture in Andhra Pradesh and Karnataka.

OBJECTIVES OF THE STUDY

The above review reveal that the studies made so far do not highlight adequately the prospects of sericulture development in the drought-prone districts of Andhra Pradesh and Karnataka. Therefore the present under taking intends to focus attention on them.

The study has the following objectives:

1) to study the progress of sericulture in India in general and in Andhra Pradesh and Karnataka in particular;

2) to analyse the economics and prospects of sericulture in selected areas of Andhra Pradesh and Karnataka States, with special reference to mulberry cultivation, silkworm rearing and cocoon production;

3) to undertake an in-depth study of employment potentialities of mulberry cultivation and silkworm rearing in the drought prone districts of Andhra Pradesh and Karnataka; and

4) to highlight the problems associated with mulberry cultivation and silkworm rearing.

HYPOTHESES

The following specific hypotheses are formulated to meet the above objectives of the study:

1) taking up sericulture enables small and marginal farmers to become economically viable;

2) mulberry cultivation and silkworm rearing generates more employment opportunities for family labour as well as hired labour; and

3) initial investment for mulberry cultivation and silkworm rearing per acre is very high.

METHODOLOGY

Sericulture is now practiced in almost all the districts of Andhra Pradesh and Karnataka. It is not possible to cover all the districts in this inquiry. To keep the study within manageable limits without in any way minimising its significance, two districts, one from each state, Anantapur from Andhra Pradesh and Kolar from Karnataka - have been selected as the study area. These districts are leading districts in mulberry raw silk production in their respective states and both are drought-prone districts. And in them mulberry is grown under irrigated conditions. Within in the study area, six mandals/taluks have been selected at the rate of three mandals/taluks from each district for sample survey. Kolar, Chintamani and Siddlagatta taluks in the Kolar district and Hindupur, Madakasira and Kadiri mandals in the Anantapur district where sericulture is taken up very intensively are selected. Again six villages, one from each selected taluk/mandal viz., Suguturu from Kolar, Doddganjur from Chintamani, Sonnanahalli from Sidlaghatta Rachepalle from Hindupur, Papasanipalli from Madakasira and Mothukapalle from Kadiri have been drawn for the sample survey.

Secondary data and information are collected from the Annual Reports of the Department of Sericulture, Andhra Pradesh and Karnataka and also from, the District Statistical Year Books from the Chief Planning Officers of the Anantapur and Kolar districts. The studies and reports brought out by the Central Silk Board, Institute for Social and Economic Change, Bangalore, Ministry of Textiles and its constituents and Central Sericultural Research and Training Institute, Mysore are also drawn upon for secondary data.

The primary data is collected with the help of a structured schedule from 180 sample farmers in the two selected districts, drawn at the rate of ninety from each district, by using the multistage random sampling method. The ninety sample farmers are drawn at the rate of thirty farmers from each of the three villages selected from each district. Of the thirty farmers taken for the study from each village, twenty are sericulturists and ten are non-sericulturists, thus making the total sample comprise 60 siriculturists and 30 non-sericulturists in each district.

The total sample of 180 farmers includes 44 marginal, 65 small and 71 big farmers. The sample also covers 36 Scheduled Caste/Scheduled Tribe farmers and 28 Backward Caste farmers.

In a majority of cases accurate information about the production of mulberry leaf and the sale proceeds of cocoons was not available from respondents as most of them did not maintain any records.

As the number of mulberry crops raised in a year varies from year to year, it is difficult to compare the net returns from sericulture over a period of time with the other principal crops raised by the farmers engaged in sericulture.

STATISTICAL TOOLS USED

The following statistical tools are used to test the hypotheses of the study.

COMPOUND GROWTH RATE

To analyse the growth of area under mulberry cultivation, cocoon and raw silk production and quantity of çocoon transacted in Andhra Pradesh and Kanataka with particular reference to the Anantapur and Kolar districts during the period 1980-81 to 1993-94, the following equation is used

y = ßt
log y = log + t log ß
Where y = Area under Mulberry / Cocoon Production
Raw silk Production / Quantity of Concoons Transacted
t = Time Period (in Years)
a = Intercept
ß = Regression Coefficient
Compound Growth Rate = (beta HAT - 1) X 100

To test the significance of Growth Rates the following t - test has been used

t~=~{beta HAT } over {SE (beta hat)}~SIM~t SUB {n-2}
where

SE~ ~(beta hat)~=~sqrt {{S sup 2} over {SIGMA ~x sup 2}}
where

S sup 2~=~{SIGMA ~e sup 2} over {n-k}

SE = Standard Error

C. STUDENT 't' TEST

In order to test the significance of the difference between the mean values of employment generation through sericulture and other crops the student 't' test has been used with the following formula t~=~ {x bar sub 1~--~x bar sub 2} over sqrt{ S sub 1 sup 2~sub /n sub 1 + S sub 2 sup 2 /n sub 2}

x bar sub 1 = Mean man days created by the adoption of sericulture.

x bar sub 2 = Mean man days created by the other principal crops (other than sericulture)

n_1 = Number of sericulturists in Anantapur / Kolar.

n_2 = Number of non-sericulturists in Anantapur / Kolar.

D. PRINCIPAL COMPONENT ANALYSIS

Principal Component analysis is a method for separating the total variation of variables into Orthogonal sub-set. The Principal Components can also be extracted from a series of random variables and thus the technique is applicable even to dependent variable included in the economic models.

Principal component Analysis is a method for reducing correlated measurement variables (p) to a smaller set of statistically independent linear combinations having certain uique properties with regard to characterizing individual differences (Overall and Klett, 1972). The first principal component is that weighted combination of the several original variables which accounts for a maximum amount of the total variation or individual differences, represented in the complete set of original variables. The second principal component is that weighted combination of several original variables uncorrelated with the first principal component and which accounts for a maximum amount of the remaining variation or individual difference similarly the rth principal component is that weighted combination independent of the first (r-1) and which accounts for a maximum amount of the remaining variation among individuals in terms of their original score values. The properties of statistical Orthogonality (independence) and maximization of variance uniquely define the principal components.

Definition : If 'y' is a linear combination of 'p' correlated measurements obtained by multiplying the original measurements 'x' (in deviation form) by a vector of weighing co-efficient 'a' the variance of 'y' over 'n' individuals can be calculated directed from the co-variance or correlation matrix of 'p' measurements by pre and post - multiplying by the vector of weighing co-efficient i.e., $V(y) = a^1 Ma$ where V(y) denotes the variance of 'y' over 'n' individuals.

The first principal component is defined as the vector of weighing Co-efficient of which maximizes variance V(y) subject to the restriction that the sum of squares for the a' equals a constant, which is conventionally taken as $a'^2 = 1$. The function to be maximized is thus

$$f(a) = \frac{a'Ma}{a'a} \quad \text{.........(1)}$$

Maximizing this function with regard to the ai yields the following matrix equation, in which λ is a Log range multiplier which is also called the characteristic root of the matrix equation.

$$(M - \lambda_1)a \text{ and } a'a = 1 \quad \text{........... (2)}$$

The vector 'a' is the directed vector of weighing Co-efficient which will maximize the criterionfunction f(a) or in other words, principal component vector. If a correlation matrix is analysed by the method of principal components the elements of the vector 'a' are weighing Co-efficient to be applied to the original variables in a standard - score form. If a co-variance matrix is used, the weighing Co-efficient in 'a' are applied to variables in a raw - score form

$$(M - \lambda_1)a = Ma - \lambda a \text{............. (3)}$$

The variance associated with a principal component is the multiplier, given the restriction that a' a = 1.

A unique property of principal components is that successive principal component vectors possess both geometric and statistical orthogonality.

$$a^{(i)} M a^{(j)} = a^{(i)} a^{(j)} = 0 \qquad (4)$$

and $A' A = I$

The present analysis is carried out on a set of variables. The variables included are:

X_1 = Area under irrigation
X_2 = Area under mulberry crop
X_3 = Size of landholdings
X_4 = Cropping intensity

X_5 = Educational levels of the respondents Ranks given :

(a) Illiterates = 1
(b) Primary = 2
(c) Secondary = 3
(d) College = 4

X_6 = Human labour (Owned)
X_7 = Human labour (Hired)
X_8 = Irrigation Charges
X_9 = Farm yard manure
X_{10} = Chemical fertilizer
X_{11} = Disinfactants
X_{12} = Quantity of mulberry leaves fed
X_{13} = Income from live - stock
X_{14} = Number of layings
X_{15} = Lighting charges
X_{16} = Cost of production of other crops
X_{17} = Returns from silk Cocoons
X_{18} = Income from other crops

The analysis is carried out separately for sericulturists in both the Anantapur and Kolar districts.

E. LOG LINEAR FUNCTION

To study the functional relationship between dependent and independent variables, the Log linear function has been used.

The function employed in the study can be represented as

Y-=-beta sub 0-X sub 1 sup beta sub 1-X sub 2 sup beta sub 2-X sub 3 sup beta sub 3-........ X sub n sup beta sub n-e sup u

where y = Dependent variable
X_1, X_2, X_3 ... and X_n are independent variables.
βi's (i = 1,2,3, ... n) = Regression Co-efficients
β_o = Intercept
u = Error term
In logaritham form it would be
$\log y = \log \beta_o + \beta_1 \log X_1 + \beta_2 \log X_2 + ... + \beta_n \log X_n + u$

The function is fitted for mulberry leaves production and cocoon production. The variables included in each of these productions are indicated below :

A. MULBERRY LEAF PRODUCTION

Y = Gross output (yield of mulberry leaves in kgs)

X_1 = Mulberry area in acres

X_2 = Human labour in man days

X_3 = Cost of manures and fertilizers in rupees

X_4 = Irrigation charges in rupees

X_5 = Share of establishment charges in rupees

B. SILK COCOON PRODUCTION

Y = Gross output (yield of Cocoons in kgs)

X_1 = Mulberry leaves in kgs

X_2 = Disease free layings in number

X_3 = Human labour in man days

X_4 = Cost of disinfectants in rupees.

X_5 = Lighting charges in rupees

Apart from the above tools simple techniques like calculation of percentages ratios and mean values are also used.

CHAPTER SCHEME

The entire study is divided into nine chapters including the present chapter and a brief description of each chapter is presented here.

In the first chapter which is introductory, the importance of sericulture in Indian economy and its suitability has been outlined. The need for the present study, review of relevant literature, the methodology adopted, and the limitations of the study have been pointed out.

The second chapter is concerned with explaining the significance of sericulture in India's rural economy. The historical background of sericulture, types of silks, the process of silk production, advantages of sericulture, its employment potential and the role it plays in the development of rural areas, are dealt with.

The third chapter presents the status of sericulture in Southeast Asian countries like China, Japan, Korea, etc and a few western countries like Italy, France, Germany, Brazil etc.

The fourth chapter analyses the progress of sericulture in India. It provides a detailed account of the emergence and expansion of sericulture in India, its progress with regard to area under mulberry, cocoon and raw silk production, employment generation etc. It also covers various plans and schemes which have been implemented in the country for the development of sericulture and the present status of sericulture in different states.

The fifth chapter deals with the growth of sericulture in Andhra Pradesh and Karnataka. It gives a detailed account of the area under mulberry, cocoon and raw silk production, marketing, infrastructural facilities, the details of NSP and other plans and programmes under implementation in these two states.

The sixth chapter deals with an empirical account of the economics and prospects of sericulture in the Anantapur and Kolar districts. It has three sections. The first section of this chapter presents the development of sericulture in these districts. The second section covers the socio-economic status of the sample farmers, both sericulturists and non-sericulturists. The third section deals with a comparison of the costs and returns of the sericulture enterprise with those of other crops.

The seventh chapter presents a clear picture of the employment generation in both mulberry cultivation and silkworm rearing. It also covers the employment potential of sericulture in comparison with that of other crops like paddy, groundnut and ragi in the sample districts.

The eighth chapter deals with the problems faced by the sericulturists. It highlights the problems of sericulturists pertaining to mulberry cultivation, silkworm rearing, extension, fluctuations in the cocoon prices, marketing etc.

The last chapter presents a summary of the findings of the present study and its conclusions.

Notes:-

1. Vasant Desai, A Study of Rural Economics, Himalaya Publishers, New Delhi, 1990, p.3.
2. World Bank, World Development - Report, 1993, p.240.
3. The *Economic Survey 1993-94,* Ministry of Finance, Economic Division, Government of India, p.148.
4. Census of India 1991, Series - I, p.12.
5. Government of India, Planning Commission, *First Five Year Plan - A draft outline,* New Delhi, 1957, p.19.
6. G.V.K. Rao and R. Thamarajkshi, "Some Aspects of Growth in Indian Agriculture", *Economic and Political Weekly,* Vol. XIII, Nos. 51-52, December, 23-30, 1978, p.18.
7. J.N. Sinha, "Rural Employment Planning - Dimensions and Constraints", *Economic and Political Weekly,* Vol. XIII, Nos. 6&7, Annual Number, 1978.
8. V.Y. Vyas and G. Mathai, "Farm and Non-farm Employment in Rural Areas - A Perspective for Planning", *Economic and Political Weekly,* Vol. XIII, Nos. 6&7, Annual Number 1978.
9. S.K. Rao and Amal Snyal, "On Promoting Employment through Labour Intensity of Techniques", *Economic and Political Weekly,* Vol. XIII, No. 6&7, Annual Number 1978.
10. M.L. Dantwala, E., "Rural Employment Facts and Issues", *Economic and Political weekly,* Vol. XIV, No. 2, June 23, 1979.
11. Director, Department of Sericulture, Bangalore.
12. Director, Department of Sericulture, Hyderabad.

CHAPTER 2
Significance of Sericulture in the Rural Economy

Today sericulture shows a picture of a well-knit agro-based, labour intensive, activity which plays a significant role in the development of rural economy. Although, agriculture remains the main source of income, directly or indirectly, for a majority of the people in rural areas, most of the rural agricultural farmers are small and marginal. It is realised that these small and marginal farmers constitute a resource base for major agricultural programmes so that increase in productivity for them can provide a significant boost to the national income distribution.

Sericulture being an agro-based rural industry, is highly suitable to provide employment to the farmers and landless labourers. It is mainly a rural and labour intensive industry requiring relatively low investment and having high potential for profit and foreign exchange earnings.[1]

As an agro-based industry, sericulture has two main aspects. The first is an agricultural aspect in which in mulberry is cultivated and silkworms are reared, and the second is an industrial aspect in which the cocoons are reeled, silk- processed, woven into fabric, dyed and printed. So the final product of sericulture is silk which always has demand in the global market.[2]

ORIGIN OF SERICULTURE AND SILK

There are many theories about the origin of silk which is regarded as the, 'Queen of Textiles'. But the most accepted theory is that it was discovered in China. Legend attributes the unravelling of silk's secret to the Chinese Empress 'XI, Si Ling Shi', wife of 'Haung Ti' (2640 B.C.). When she was sitting under a mulberry tree with a cup of hot tea before her, a silk cocoon from the tree accidentally fell into the cup. In her attempts to take out the cocoon, lustrous strands of silk came into her fingers, which were slowly unwinding. Attracted by this glittering thread, the Empress gave it a lot of attention and made silk a fashionable cloth. The Chinese named this lustrous cloth "Si" after their Empress which is still the Chinese word for silk.[3]

In China, the birth place of silk and silk-weaving art, the fibre was so treasured that it became a measure of currency and reward. The imperial court established gifts to foreign powers thereby extending influence over its neighbours. They guarded the secret of silk for hundreds of years and displayed its beauty to foreign visitors and fancied it as 'The Hair of Sea-Sheep'. The imperial law decreed death by torture to those who disclosed the secret of silk. But as the empire declined, the circle of those who knew of the illustrious and sensual fiber widened.

Silk industry was brought to India from China through the famous 'Silk Road' via Khotan in 140 B.C.[4] Indian scholars,however, believe that sericulture was practised in the foot hills of the Sub-Himalayas much before that date, and the silk culture originated somewhere in the areas flanking the Ganga and the Bramhaputra. But the ancient Indian Sanskrit text the 'Amarakosha' confirms that the origin of silk industry was in China. Silk was called 'Cheenamsokam ' or 'Cloth of China', in some of the older Sanskrit works.[5] It is believed that silkworm eggs and mulberry plants were carried to India by a Chinese Princess from a province near Indian border.

In a strict sense, the term 'sericulture' refers to the process involved in the production of natural silk. 'Serio' is a Latin word which means silk. Silk is a natural filament created by the silkworm. Therefore, sericulture means the raising or rearing of silkworms for the production of silk. The activity gets commercialised and developed into a largescale endeavour with specialised activities concerned with the rearing of silkworms. Hence, sericulture would also refer to a composite activity including the raising of mulberry which is the food plant for silkworms, production of silkworm eggs or layings, silkworm rearing and ending with the disposal of silk worms.

TYPES OF SILK

Silk is a fibrous protein of animal origin. A number of animals secrete silk which is used by them for anchorage (mussels), to entangle their prey (spiders), or to form a protective sheath with or without other materials. Nearly 400-500 species are known to produce silk but only a very few are commercially exploited. At present nearly 95 per cent of the commercial silk is coming from mulberry silkworms called 'Bombyx mori', Commercial silks made from all other sources are collectively called non-mulberry silks as mulberry is not the food-plant for the worms producing them.

Depending upon the race, species of the worm and the food plant on which they feed silks are differentiated into four kinds : 1) Tasar or

Tussar Silk ; 2) Eri Silk ; 3) Muga Silk and 4) Mulberry Silk . The first three types of silks are also called wild silks since the worms producing them are not fully domesticated, unlike the mulberry worm, and are practised in forests and hilly areas.

TASAR SILK (TUSSAR SILK OR TUSSORE SILK)

Tasar silk is produced by a species called 'Antheeraca Mylitta'. This exists in the form of nearly 19 ecotypes thriving between 450 and 750 metres above Mean Sea Level (MSL). It is a wild species found in uni, bi and trivoltines. It is polyphagous in nature. The main food plants of Tasar silk worms are, Assan, Arjun and Sal. Tasar silk is copperish coloured and does not possess the lusture of mulberry silk. A new strain of tasar worm has been developed from 'Antherea Royelie' and 'Antherea Pernyi' which feed on oak trees abounding at 600 - 1800m above MSL. Oak tasar silk is yellow to white in colour, and it is the finest among the non-mulberry silks. Tasar culture is found in Bihar, Orissa, Madhya Pradesh, Karnataka and Andhra Pradesh.[6]

ERI SILK

Eri Silk is also known as 'Endi' and ranks next to Tasar in commercial importance. The genus 'Philosamia' comprises 17 species distributed mainly in the Indo-Australian region. The silk worm 'Phelosamia Rivoinie' thrives in India and is multivoltine, yielding four to six crops a year. The food of Eri SilkWorm is castor leaves, Papaya and Payam. Eri cocoons are white and brick red in colour. Eri-culture is found in Orissa, Bihar, West Bengal, Madhya Pradesh, Karnataka and Andhra Pradesh.[7]

MUGA SILK

Muga silk is obtained from the worms 'Antherea Asamensis' which are polyphagous, primarily feeding on Som and Salu. The silk is golden yellow to creamy white in colour with a lustrous look. Muga silk is almost the monopoly of Assam, but it is also found in Meghalaya.[8]

MULBERRY SILK

What is commonly called silk is mulberry silk. 'Bombyx Mori', a native of Northern China or Bengal, is the silkworm which produces mulberry silk. It is found in all voltines. Mulberry sericulture is being practised in almost all the states of India. The production of mulberry silk involves a series of processes. As it predominates over other types of silk, here a special mention has been made of the processes of its production. The processes for producing other kinds of silk are broadly similar, but there are important differences in the suitability of food plants, the choice of rearing conditions, and the appropriateness of reeling techniques. Even the process of mulberry sericulture itself differs from one region to another depending upon various aspects like the geographical features, the flora and fauna of the area where it is practised. Sometimes technological advances also determine the practice of sericulture.

In countries like Japan, Russia, China and Korea, silkworm rearing is not possible throughout the year since the winter and cold months in those countries are for a long duration and affect the worm's activity. In the warm regions of India, sericulture is a perennial activity and silkworm rearing can be carried throughout the year as the variations in climate are not so extreme. The practice of mulberry sericulture in Andhra Pradesh, Karnataka and Tamil Nadu is very similar. A description of the practices adopted in these states is given in the following pages.

Mulberry sericulture has two broad operations, namely 1) Mulberry cultivation and 2) Silkworm rearing.

MULBERRY CULTIVATION

It is purely an agricultural operation and a major factor which determines the quantity and quality of the production of the mulberry leaves. The production of mulberry leaves is a labour intensive operation. This concious cultivation of mulberry plants for harvesting their leaves, the food for silkworms, is referred to as 'moriculture' which plays a key role in the economics of commercial sericulture.[9] The returns from sericulture, namely quality cocoons, depend mainly on the nutritive value of mulberry leaves supplied to the worms during rearing. Though as many as eleven plants have been suggested as alternate food plants for 'Bombyx mori', the larval life of the silkworm is prolonged by them and the cocoons produced are small in size. Therefore, commercial sericulture still depends largely on mulberry plants.

The main objective of mulberry cultivation is to obtain large quantities of highly nutritive leaves. This is influenced by many factors like climatic conditions, nature of soils, variety of the mulberry plant, initial establishment of mulberry garden and proper management of it.

Mulberry is a perennial crop and it continuously and consistently yields leaves for more than 15-20 years. Hence there should be no hesitation regarding the cost of initial establishment, and it should be carried out on scientific and systematic lines. There are a number of mulberry varieties and forms belonging to different species and sub-species. Mulberry can be propagated both by cuttings and saplings. The mulberry plant yields leaf within four or five months after plantation under irrigated conditions. The leaf yield of mulberry plantation varies enormously depending on the inputs. It ranges from 3000 to 10,000 kgs, under rainfed conditions, and from 10,000 to 40,000 Kgs under irrigated conditions per hectare.[10]

SILKWORM REARING

Silkworm rearing is a cottage activity as distinct from mulberry cultivation. The sericulturist rears silkworms either in his own house or in a separate shed built for this purpose. This activity needs much manual attention and skill. There are four silkworm races viz., 'univoltine', 'bivoltine', 'trivoltine' and 'multi-voltine'. In 'the univoltine', the silk worm passes through only one cycle and remains dormant in the egg stage or in the pupa stage in the cocoon for a long time; that is, it has only one breed or generation within a year. The 'multivoltine', has cycles after cycles, where a cycle just takes only six weeks. The 'univoltines' are very sensitive to high temparature and are to be reared in cooler regions. The cocoons are bigger in size and posses more silk. The 'multivoltines' are suitable to warm regions and many rearings can be had in a year. But the size of the cocoon is smaller and its quality is low. The 'Bivoltines' and 'Trivoltines', fall in between these two.

The silkworm undergoes four distinct stages in its life cycle : egg, larva, pupa and moth. The moth lays eggs and from these eggs tiny ant-like worms hatch out and feed on mulberry leaves. This is the second stage called the larva or caterpillar. The larva when fully grown spins silk into a cocoon around itself. Inside the cocoon the worm is transformed into the pupa which is the third stage. After some days the pupa develops into a moth, which is the final stage. The moth comes out of the cocoon and after couplation the female moth lays eggs. This life cycle is thus repeated. The length of the cycle differs from region to region and from one species to another depending on factors like climate, vegetation and temperature.

The sericulturist purchases eggs or disease-free layings (dfls) or industrial seeds generally from government grainages or licenced seed-producers. In each laying there would be 350 to 400 eggs. The eggs are white or whitish yellow in colour. They hatch within nine or ten days after their laying. Tiny ant-like worms come out of the shells. Each larva is a few millimetres long and weighs about 0.5 mg. Upto this, the process is called 'Chawkie Rearing'. The mulberry leaves are cut into very fine pieces and sprinkled over the larvae, the which are fed for about 24 to 25 days. As the worms grow they are fed with bigger and bigger leaf bits.

During the larva stage the silk worm moults four times. Moulting is a process where the silkworm casts of its existing skin in place of a new one. The period before or after moulting is called 'instar'. There are five instars in a silk- worm's life. During IV and V instars intake of mulberry is very high. By the end of V instar the caterpillar increases its weight 10,000 times, showing a weight of four to five grams. Proper care should be taken during IV instar since its duration is greater than that of the first three. If takes nearly 30 hours under optimum conditions of temperature and humidity.

The larvae are fed with leaf during the early ages and with both leaf and shoot during the later stages. The rearing bed spacing should be 0.2 Sq.m for 50 dfls in the beginning.[11] During IV and V instars the bed area should be increased in proportion to the growth of the silkworm larvae. Bed cleaning should be done with nylon or cotton nets of appropriate mesh size. Heavy crop losses are recorded due to infections - 'Flacherie', 'Bacterial Flacherie', 'Nuclear Polyhedrosis' etc., 'Pebrine', 'Muscardine' also occur occassionally. Recently crops were damaged due to the 'uzy fly'. To prevent the crop damage due to the 'uzy fly', silkworm rearing under nylon nets has become compulsory.

The fully grown silkworms are kept in a proper frame to facilitate the spinning of cocoons. Such frames are called 'mountages'. During the fifth instar, the larvae become ripe for spinning between the fifth and seventh day. They should be picked up and placed on mountages in right time to avoid silk wastage. Worms premature or too late after ripening can give poor quality cocoons. Moreover, humidity, ventilation and temperature also influence the cocoon quality. This process is called 'mounting' or cocoon production.

Mountages used in India range from simple dried grass and twigs in Assam, and dried mulberry twigs in Jammu and Kashmir to bamboo spirals popularly called 'Chandrikas' in all the southern states and West Bengal. Mounting and harvesting are difficult in the first two types as they require more labour and also results in more double and deformed cocoons. The bamboo mountage with 12 mm. holes in the back of the mat is found the best with least double and urinated cocoons, higher reelability and better raw silk recovery. 'Plastic Collapsible mountage' for its convenience of handling and storage and 'Bottle bursh mountage' for its low cost are also popular.[12]

The ripe worms must be mounted at the rate of 440 to 660 worms per square meter depending upon the silkworm variety. This helps to obtain good cocoon build, lesser double cocoon and lesser urinated cocoons.

The spinning of cocoons is completed in two to three days. The worm inside the cocoon turns into the pupa on the fourth or fifth day depending upon silkworm variety. On this day the pupa is hard and the cocoon shell is dry and suitable for long distance transportation for marketing. The Cocoon yield will be 40 to 45 kg. in multivoltine and 50 kg. in bivoltine hybrid for 100 layings. It is said that a silkworm can yield 800m length of yarn during its lifetime.

The Cocoon yield depends on the production of quality silkworm eggs. The seed organisation plays a vital role on which the success of the industry depends. The layings are called disease-free layings (dfls). Hence care has to be taken that the layings are healthy and made sure that they are free from any disease or infection after microscopic tests. The establishment where the layings are produced is called grainage. Generally rearers use cross-breed layings produced from a multivoltine male Local Race (LR) and a bivoltine female Foreign Race (FR). The FR and LR dfls are also called basic seed and are produced by government grainages in their respective grainages. These dfls are supplied to notified rearers for further multiplication. These rearers breed the dfls and sell the resultant seed cocoons to the government grainages. In the cross-breed grainages the commercial seed is produced by cross-breeding.

SERICULTURE AND INDIAN ECONOMY

Sericulture occupies a unique position in Indian economy. It is an important and highly rewarding occupation among the developing countries. India enjoys a very favourable position in the production of mulberry raw silk. Its inherent qualities like, high profit and labour intensity, have motivated the policy-makers, administrators and scientists to propogate the sericulture enterprise throughout the country as a suitable answer to rural unemployment, and low per capita income.

It has been found that sericulture is highly profitable as compared to many other crops. Owing to many advantages mulberry has been replacing other important crops in many regions of India, for example sugarcane in karnataka, groundnut in Andhra Pradesh, jute in West Bengal etc.

ADVANTAGES OF SERICULTURE

Till about a few years ago sericulture was considered as a subsidiary occupation of poor farmers, while the big landlords used to cultivate only food crops, fruits, orchards, etc. Now sericulture is no longer a subsidiary occupation and more and more farmers including big landlords are taking it up. The various advantages of sericulture are as follows.

Sericulture is a labour intensive rural industry capable of creating employment to around 12 persons per hectare of mulberry. It requires low gestation period and it starts yielding just in five or six months after raising the mulberry plantation. Once the plantation is established it requires low investment and it will continue to yield for 15 to 20 years and give higher returns with a little expenditure on maintenance. It involves simple technologies which are easy to understand and adopt even by illiterate farmers. It gives returns in quick succession yielding income for every two or three months. Sericulture does not involve hard labour and therefore the rearing of silkworms is generally attended to by women and old people. It does not require sophisticated machinery and it involves the use of simple appliances. The mulberry plant can be grown on any type of soil, even on forest fringes and hill slopes. It withstands severe drought conditions and yields at least some income for sustenance, while other agricultural crops wither away under similar conditions. Sericulture suits even small and marginal farmers owning less than 0.50 hectare of mulberry plantation. Therefore it could be an advantageous industry for improving the economy of the retarted sectors of society, like the Scheduled Castes and the Scheduled Tribes. Nothing goes waste in the sericulture industry. The bi-products, like mulberry twigs, silk worm litter, pupae, pierced and unreeliable cocoons, reeling waste, etc., are all useful in one form or the other and bring income. Raw silk has very good demand both in the domestic and international markets.[13]

EMPLOYMENT GENERATION

At present, sericulture provides employment to around six million persons in India and most of them belonging to the weaker sections of society. It can be said that no other single activity, whether agricultural or industrial, can bring such fortune to the most needy.

A farm labourer is able to find employment all the year round by combining mulberry cultivation with silkworm rearing. Since Indian agriculture is seasonal in character attention should be given to this aspect of employment in agriculture. In controlling unemployment and seasonal unemployment, sericulture can play an efficient and a relevant role.

The nature of employment in sericulture industry can be divided into two types; 1). Agricultural - includes mulberry cultivation, silkworm rearing, Cocoon production and 2). Industrial - includes reeling of cocoons, twisting, warping, dyeing and weaving. The former activities are rural in nature and the latter are semi-urban and urban.

As a labour intensive export-oriented cottage industry sericulture can generate, high employment and income per unit area of land. One hectare of mulberry plantation under irrigation generates about 13 work years annually in mulberry cultivation, silkworm rearing, silk reeling, twisting and weavings.[14]

The labour requirement in sericulture is more than two and half times that of paddy and ten times that of groundnut. Thus the employment generation in sericulture is very high over that of the other crops, including paddy and groundnut which are taken as double crops in a year.

INVOLVEMENT OF FAMILY LABOUR

The utilisation of family labour reduces the production cost considerably and increases the profit substantially. In the Rayalaseema Region of Andhra Pradesh the participation of family labour in mulberry cultivation and silkworm rearing is worked out to be around 68 percent.[15] However, the Central Sericultural Research and Training Institute, Mysore, in its study of "Social and economic changes among sericulturists in 1983-84" found that the share of family labour in the production of mulberry and cocoon is gradually decreasing.[16]

WOMEN IN SERICULTURE

Employment opportunities for women are also high in sericulture activities. Various operations in the production of silk beneficially engage women. Sericulture, because of its unique nature of work, proves to be an ideal activity for women who can work in addition to their regular tasks of taking care of the family. Its operations do not require hard labour. Almost all the sericulture activities, except such tasks as digging, ploughing and carrying heavy loads, which are strenuous can be carried out by women independently. Silk- worms being delicate have to be handled with proper care. Thus the entire process of rearing needs expertise, high skill and patience. Women possess these qualities to an eminent degree and therefore are more suitable than men. It is worked out that about 2,760 women work days comprising about 60 percent are generated per annum out of a total of about 4,631 work days in all the activities in sericulture per hectare of irrigated mulberry[17]. Thus sericulture provides scope for the direct involment of women in the process of production and decision making, for improving their economic conditiions and for giving them greater recognition and status in the family and society. Under the National Sericulture Project, the action plan on women envisages group-formation, special training programmes for women, allotment of land in the names of women, special credit schemes for women etc. All these programmes are under implementation for raising the active participation of women in sericulture.

INDIRECT EMPLOYMENT EFFECTS

The whole area of raw silk industry in the broadest sense will also provide new employment. In silk reeling activity there is a considerable scope for generation of employment opportunities to the artisans and unskilled workers of rural areas. A reeler, with a unit of 10 basins, provides employment, to 21 persons including the reeler.[18] The organisation of reeling and silk weaving can also help some of the markets and most vulnerable sections of society in rural areas. Silk waste and pupae, the by-products of silk industry, open yet another sector of economic activities keeping people busy in processing and spinning of silk waste, producing pupae oil and pupae meal.

The mulberry cultivation, which requires more and better silkworm races as well as increased production of mulberry seed, should be a source of considerable indirect stimulous to employment. If the job opportunities created in the constructiion of irrigation wells, rearing houses, repair and replacement of rearing equipment, reeling, marketing, transport, extension, research, etc., with all their forward and backward linkages are also considered, the employment potential of sericulture will be seen to be enormous.

HIGH INCOME GENERATOR

Sericulture is a highly remunerative cash crop with rich dividents. It is the only cash crop which provides frequent and attractive returns in the tropical states of the country throughout the year. The average annual income per hectare would be around Rs.40,000/- per hectare which is substantially high when compared to that of the other tropical crops. Sericulture industry is, therefore, well suited to the small and marginal farmers who are the below poverty line. Another plus point is that the plantation takes deep roots and survives in severe drought conditions. As a drought resistant plant, mulberry is highly preferred in drought affected areas.

The silk industry plays a vital role in transferring wealth from the richer sections of society to the poorer sections. Silk is consumed mostly by the affluent and the money spent by them on the purchase of silk is distributed among the sericulturists, reelers, twisters, weavers and traders. The details of the distribution of income from the sale of soft silk fabric among different sections of society are shown below in Table 2.1. and depicted in diagram 2.1. From the table 2.1 it may be seen that the cocoon producer receives nearly 52 to 57 percent of the money from the sale of silk fabric.

BY PRODUCTS OF SERICULTURE

In sericulture nothing goes waste. Its by-products are useful in many ways. Mulberry leaves and shoots left by the silkworms form good fodder to the cattle and increases markedly their milk yield. The stifled and much-maligned pupae are used in the preparation of dog biscuits, oil, etc. The oil and the protein powder extracted from the dead pupae can be utilized in manufacturing soaps and baking industries respectively. It forms a rich food in poultry, fishery and piggery. The silkworms' excreta can be used as manure. The rational utilization and disposal of by-products helps the sericulturists to enhance their economic gains.

Considering the above advantages of sericulture it should be an effective tool for rural development. In view of its high employment potential and remunerative income generation, sericulture is regarded as one of the important means of alleviating rural poverty and ushering in rural prosperity, and therefore, receiving due attention in the rural development programmes at the State and Central levels.[19] The National Commission on Agriculture has also rightly observed that large- scale expansion of this industry would substantially benefit the vast number of small and marginal farmers besides providing employment opportunities for millions in the rural sector.[20]

Table 2.1
DISTRIBUTION OF INCOME FROM SALE OF SOFT SILK FABRIC

Sl. No.	Category of Persons	Soft Silk Fabric of		
		40 gms/m.	50 gms/m.	60 gms/m.
1.	Cocoon Producer	51.5	54.6	56.8
2.	Reeler	6.2	6.6	6.8
3.	Twister	8.3	8.7	9.1
4.	Weaver	14.5	12.3	10.7
5.	Trader	19.5	17.8	16.6
	Total	100.00	100.00	100.00

Source : Silk in India - Statistical Biennial, Central Silk Board, Bangalore, 1988

ROLE OF SERICULTURE IN THE DEVELOPMENT OF RURAL SECTOR

It is rightly said that sericulture is a boon to rural farmers. It plays an unique role in the development of rural economy. Its uniqueness lies in the fact that its activities not only engage the rural households in the cultivation of mulberry and silkworm rearing but also encompass in their fold a whole range of reelers and weavers.

A large section of farmers with uneconomic holdings and an enormous group of ruined artisans who constitute the bulk of the non-agricultural section of the rural population together tell the tragic tale of large sections of the rural population in the India.

During the last 30 years there has been a considerable growth of industries in India, but it has hardly improved the standard of life, more especially in the rural areas. The industrial growth has been slow and lop-sided, and a few largescale industries are concentrated in the cities.

As India lives in her villages and agriculture is the backbone of her economic life, all efforts must devote to solve the rural economic problems. Agriculture in India is not a business proposition for our farmers but a way of life. But many economic forces such as the steady increase in population, decay of indigenous industries, lack of other avenues of employment and the rise in land values, have been responsible for this increasing pressure on land.

Diagram 2.1
DISTRIBUTION OF INCOME FROM SALE OF SOFT SILK FABRICS OF 60 gms/m

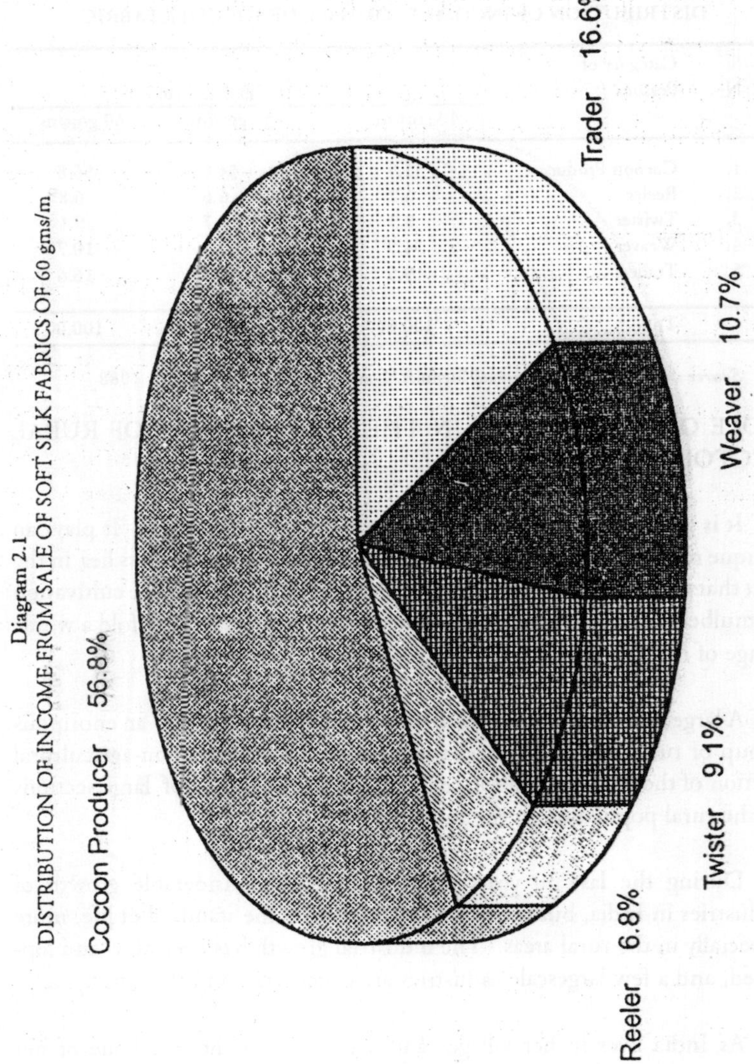

This increasing pressure of population on the soil and the growing indebtedness of the agricultural classes due to a combination of factors has led to the emergence of a class of landless labourers. These labourers may be classified into three groups: 1) field labourers - who comprise ploughmen, reapers, sowers, weeders, etc; 2) ordinary labourers - who are engaged in such works as embankment, well digging and canal silt clearing, 3) skilled labourers such as carpenters, masons, blacksmiths and leather workers and such other artisans who, though not exclusively agricultural workers, are engaged for important purposes by the farmer, and the wages paid to them are governed by those paid to the agricultural labourers in the village. It is also to be noted that in rural areas the agricultural labourer is not only unemployed over a considerable part of the year but the wages paid to him even during the period of employment are very low. The labourer's earning power is very low and in some parts of India his extreme indebtedness has driven him to work as a bonded-labourer in the fields of his creditor.

India has been an agricultural country in the past and will continue to be so in the future, although its industrial development has been quite spectacular in recent years. According to the estimates by 2001 A.D. the population would reach 986 million. Then the pressure of population on land would increase enormously resulting is further lowering of standards in agricultural production. Excessive pressure on land leads to high rate of poverty, disguised unemployment, under-employment, etc.[21]

ROLE OF SERICULTURE IN RURAL DEVELOPMENT

As noted already sericulture has a great potential to alleviate rural poverty and unemployment. Mulberry, as a drought resistant plant, could be grown even in areas with low rainfall. It has wider scope for augmenting family income and employment opportunities for small and marginal farmers. It would create employment opportunities all through the year on farm and off-farm ranging from cultivation of mulberry and rearing of silkworms to allied activities.

Compared to other crops, the work participation rate is quite high in sericulture. It also absorbs many non-cultivators in the process of picking of leaves, rearing of silkworms and other allied activities. The migration of unskilled labour is a common feature in rural areas. But sericulture not only provides income for sericulturists but also retains people from mass migration to other places in search of employment.

Sericulture could be taken up in all kinds of regions. Appropriate technologies are now available for growing mulberry under all kinds of agro-climatic conditions. In fact, in some of the traditional rainfed sericulture areas where even millets could grow only with difficulty, the rearing of local race silkworms with local variety of mulberry has provided a survival strategy for generations of small and marginal farmer households.

The highly labour intensive nature of sericulture makes it particularly suitable for the marginal and small economic groups to take it up on a sustainable basis. Rearing of silkworms is always more profitable with family labour than with hired labour. Wage workers prefer to work in silkworm rearing rather than in agricultural farm work. In reeling, the work-load is heavy and the wage rates are also higher than in any other farm work or other comparable industries. Thus sericulture provides an opportunity to the poorer sections of society to participate in the process of economic development.

Sericulture provides a means of regular and stable income with a high annual turnover of three to four crops in rainfed conditions and five to six crops under irrigated conditions. Division of the garden into plots with alternate harvest timings, could enable the household to carry on silkworm rearing throughout the year continuously. Thus a regular and steady income from sericulture helps the poor to improve their standard of living.

CONCLUSION

Sericulture, is a highly suitable agro-based rural activity which provides gainful employment to a majority of the farmers as well as landless labourers of the rural areas. Sericulture is a blend of two distinct aspects i.e. agricultural and industrial. The final product of sericulture is silk which has enormous demand in the international market and acts as means of foreign exchange earnings.

Though there are many theories about the origin of silk, historical evidence prove that China is the original home of sericulture. It was brought to India by a Chinese princess from a province near the Indian border. Now India enjoys the unique distinction of producing all the four varieties of natural silk i.e. tasar, eri, muga and mulberry and takes the second rank among the mulberry silk producing countries of the world.

Sericulture generates high, employment and income per unit of land. It creates about 13 work years annually per hectare under irrigated conditions.

It generates very high income when compared to other traditional crops like paddy, groundnut etc. The nature of activities in sericulture are such that it facilitates active involvement of women and other family members.

In view of the high employment potential and quick returns, sericulture is being considered to be one of the important and effective tools of alleviating rural poverty and ushering rural prosperity. Therefore, it is receiving due attention in the rural development programmes both at the state and the central government levels.

Notes:-

1. Dr. R.K. Datta and Dr. C. Ravikumar, "Sericulture and Rural Development", **Proceedings of the International Congress on Tropical Sericulture Practices, February, 18-23, 1988,** Central Silk Board, Bangalore, 1991, pp.1-2.
2. Dr. C. Ravikumar and Mrs. Prabha Sekhar, "Contribution of Sericulture for Rural Development", **Proceedings of the International Congress on Tropical Sericulture Practices,** Op.cit., p.80.
3. The World Book Encyclopedia, **Silk,** 1962, XVI,- Field Enterprises Educational Conformation, Illionoise, p.379.
4. Dr. R.K. Datta, "Indian Sericulture Past, Present, Future", **Souvenir, International Congress on Tropical Sericulture Practices 1988,** C.S.B., Bangalore, p. 217.
5. J.G. Royale, "Silk Culture in India", *Indian Silk*, Vol. XXV, No.2, June, 1986, p.4.
6. Dr. Ramana, **Economics of Sericulture and Silk Industry,** Deep & Deep Publications, New Delhi, 1987, p. 20.
7. Ibid., p.20.
8. Ibid., p.21.
9. G. Ganga and J.S. Chetty, **An Introduction to Sericulture,** Oxford & IBH Publishing Company Private Ltd., New Delhi, 1991, p.11.
10. D.V. Ramana, **Economics of Sericulture and Silk Industry in India,** Op.cit., P.16.
11. FAO: Sericulture Manual II, **Silk Worm Rearing** - Oxford & IBH Publishing Company Pvt. Ltd., New Delhi, 1993, P.67.
12. Dr. Manjeet S. Jolly, **Appropriate Sericulture Techniques,** CSRTI, Mysore, 1987, pp. 102-103.
13. S. Muniraju, "Sericulture Tool for Rural Development", **Souvenir, International Congress on Tropical Sericulture Practices,** Op.cit., pp.3-5.
14. Mrs. Prabha Sekhar and C. Ravikumar, "Role of women in Indian Sericulture", **Proceedings of the International Congress on Tropical Sericulture Practices,** Op.cit., p.65.
15. D. L. Narayana, **Economics of Sericulture in Rayalaseema,** op.cit.,p. 81
16. K.V. Benchamin and Manjeet S. Jolly, "Employment and Income Generation in the Rural areas through Sericulture", *Indian Silk,* Vol.XXVI, No.2, June 1987, p.9.
17. Mrs. Prabha Sekhar and C.Ravikumar, "Role of Women in Indian Sericultre", **Proceedings of the International Congress on Tropical Sericulture Practices,** Op.Cit., pp.68-69.
18. D.V.Ramana, **Economics of Sericulture & Silk Industry in India.,** Op.cit., P.37,
19. Venkata Narasaiah, **Sericulture in India,** Ashish Publishing House, New Delhi, 1992, pp.17-19.
20. A.R.S. Gopalchar, **Three decades of Sericulture Progress,** Central Silk Board, Bangalore, 1978, p.3.
21. C. Ravikumar and Prabha Sekhar, "Contribution of Sericulture for Rural Development in India", **Proceedings of the International Congress on Tropical Sericulture Practices,** Op.cit., pp.77-80.

CHAPTER 3
Status of Sericulture in Some Selected Countries

Silk, being a magical natural fiber, has had a facination for people all through the ages. Although its production has been rising gradually, it has not been available in large quantities.[1] With several ongoing projects that the silk producing countries have taken up recently, it was expected that the world silk production would touch 85,000 tonnes by the end of the year 1995.[2] It is estimated that at least 25 million farm workers are employed in sericulture as well as many others in the processing of silk.

Table 3.1

RAW SILK PRODUCTION IN THE WORLD (1938 - 1992)

(in thousand tonnes)

Sl. No.	Country	Year					
		1938	1978	1983	1985	1989	1992
1.	China	4.86	19.00	28.14	32.00	40.70	51.00
		(8.60)	(38.49)	(50.26)	(54.32)	(60.76)	(63.02)
2.	India	0.69	3.48	5.69	7.03	10.00	13.00
		(1.22)	(7.06)	(10.16)	(11.93)	(14.93)	(16.06)
3.	Japan	43.15	15.96	12.46	9.58	6.08	5.08
		(76.34)	(32.33)	(22.25)	(16.26)	(9.08)	(6.28)
4.	Brazil	0.04	1.24	1.36	1.46	1.90	2.30
		(0.07)	(2.51)	(2.43)	(2.48)	(2.84)	(2.84)
5.	Korea	1.83	4.24	1.94	2.09	1.20	1.38
		(3.24)	(8.59)	(3.47)	(3.55)	(1.79)	(1.70)
6.	Others	5.95	5.44	6.40	6.75	7.10	8.17
		(10.53)	(11.02)	(11.43)	(11.46)	(10.60)	(10.10)
	TOTAL	56.52	49.36	55.99	58.91	66.98	80.93
		(100.00)	(100.00)	(100.00)	(100.00)	(100.00)	(100.00)

Source : 1. T.D. Koshy, Silk Exports and Development.
2. Japan Raw Silk and Sugar Stabilization Agency.

Note : Figures in parentheses indicate computed percentages.

All the developing countries are now exploring the possibilities to take up silk production primarily due to the large number of employment opportunities it creates to the rural poor and to fetch the much needed foreign exchange to the country.

A look at the development of sericulture in the world reveals its glorious period as well as its setback period. The Table 3.1 gives the details of raw silk production in the world between 1938 and 1992.

It is evident from Table 3.1 that during 1938 the production of silk was only 56.52 thousand tonnes and it decreased to 49.36 tonnes by the end of 1978. After 1978 sericulture showed a rapid development and by the end of 1992 silk production reached 80.93 thousand tonnes. The details of output of mulberry cocoons in the world between 1990 and 1992 are presented in Table 3.2.

It is clear from the Table 3.2 that during 1990 and 1992 the mulberry cocoon production in the world has registered a substantial growth i.e. from 734.20 to 858.60 thousand tonnes.

Table 3.2

OUTPUT OF MULBERRY COCOON IN THE WORLD (1990-1992)

(in thousand tonnes)

Sl. No.	Country	Year			Percentage to total
		1990	1991	1992	(1992)
1	China	480.00	551.00	600.00	69.88
2.	India	116.70	107.20	129.70	15.10
3.	Japan	24.90	20.80	15.60	1.82
4.	C.I.A	46.00	46.00	46.00	5.36
5.	Brazil	15.80	17.20	17.60	2.05
6.	Thailand	17.40	18.00	18.00	2.10
7.	Korea	18.60	17.90	16.60	1.93
8.	Vietnam	4.50	6.00	4.80	0.56
9.	Turkey	1.30	1.30	1.30	0.15
10.	Others	9.00	9.00	9.00	1.05
	TOTAL	734.20	794.40	858.60	100.00

Source : Japan Raw Silk and Sugar Stabilization Agency.

Historical evidence shows that silk was developed in China at least 4,500 years ago. For many centuries it was a closely guarded secret, but gradually knowledge of sericulture spread both eastwards and westwards, to Japan, Korea, India, Persia and eventually Western Europe and America.

Now the industry has spread to about 58 countries and is being widely practiced in temperate countries. From Table 3.1 it may be observed that at present China is the leading producer of raw silk. Japan, once a leading producer drastically cut its production of silk and came down to the second place. By 1985 it further declined to the third place allowing India to occupy the second rank. During the year 1992 the shares of China, India, Japan and Korea were to the tune of 63.02, 16.06, 6.28 and 1.70 per cent respectively, in the total world raw silk production.

SOUTHEAST ASIAN COUNTRIES

One can easily say that sericulture enjoys rapid growth mostly in the Southeast Asian countries and in a few wesstern countries. Nearly 87 per cent of the world raw silk production is from these countries. This is largely due to the low level of their economy and the suitability of sericulture to family labour and low-paid rural labour in them. In addition, favourable weather conditions, low investment with quick profit and high market demand, all have contributed greatly to the expansion of sericulture.

CHINA

China, undoubtedly is the homeland of silk. No other country in the world has the history of silk so closely woven with it. The history of silk can be traced back to China centuries before the Christian era, where there is evidence of a written symbol for silk as early as 2600 B.C.[3]

The Post-Revolution period saw a systematic development of silk industry in China. China now controls the world silk market with a share of 63 per cent in raw silk trade. It produced 51,000 tonnes of raw silk during 1992.

It is estimated that in China 20 million households are engaged in silk industry, and out of the 30 provinces, 28 are engaged in silk production. China produces silk which is superior in quality to the silk produced in any other country in the world. About 70 per cent of the silk produced in China is bivoltine, which is considered to be the best. More than 1.2 million hectares are under mulberry cultivation and more than one million people are employed in China's silk enterprises.[4]

State organisations in China play a vital role in the production and trading process of silk, although recently there has been a substantial decentralisation in decision making. Specialised agencies have control over specific operations and this has created difficulties in achieving an overall view of developments. In 1982 the China Silk Corporation (CSC) was established to manage all activities involving production, purchase and sale of silk. Efforts have been made to integrate production and sales and has become more responsive to market requirements. As a part of this the CSC branches appear to have been given greater autonomy.

Agricultural institutes undertake silkworm egg production, maintenance of stocks of pure races, and provide sufficient quantities of disease-free layings to the farmers. Production of cocoons is organised by the communes which have production targets set by the Ministry of Agriculture. The cocoons are purchased by the CSC. Cocoon drying is undertaken at the collection centres and the CSC allots them to the reeling factories, according to its production plans.[5]

Cocoon reeling is organised both in filatures and small workshops. Within China there are substantial differences in the production technologies ranging from traditional labour intensive methods to automated capital intensive technologies. In recent years China has become more oriented towards international market requirements. Semi-automatic and multi-end reeling machines are used to produce international grades of raw silk, whilst new processes and procedures are being adopted in weaving, dyeing, printing, finishing and quality control. Sometimes foreign expertise is also used in these matters.

Conflicting opinions are expressed about the future productive trends in China. Some Japanese officials, who closely follow the developments in China, hold the view that production may come down to 2,40,000 tonnes of mulberry cocoons, citing as reasons for their view the need in that country to produce sufficient food crops, the existence of alternative profitable crops, poor soil conditions, limited fertiliser supplies, variable weather conditions and stable overseas demand.

However, there is much room for improvement in mulberry, cocoon and raw silk yields through the use of fertilizers and better mulberry and silkworm varieties, as well as improved methods of processing and management. Indeed it has been suggested that even with a stable mulberry hectarage, China could double its raw silk production in twenty years. No other country is anywhere near China in silk production both in terms of quality and quantity. China thus is poised to maintain its superiority and near monopoly of silk trade in the years to come.

JAPAN

Japan was considered to be the second home of silk. Although China gave birth to silk, it was Japan, which developed silk technology on modern lines. Japan maintained its position for centuries as the world's second largest producer of silk. However, India very recently has replaced Japan to take the second position.

The decline in Japanese silk production was largely due to (1) the labour intensive nature of the sericulture industry - the unwillingness of the younger generation to take up this as a full time profession (2) the unfavourable weather conditions of 1989 which led to a shortage of mulberry leaves and (3) the onslaught of the electronic industry which led to the downfall of the sericulture industry.

Silk production in Japan reached its peak in 1930-31 when 3,80,000 tonnes of cocoons were produced. Following the slump during the Second World War, the production revived in the 1950s and raw silk peaked again in the late 1960s averaging around 20,000 tonnes per annum. By 1983 cocoon and raw silk production fell down to 60,700 tonnes and 12,460 tonnes respectively. With rapid industrial development there has been diversion of land and labour away from sericulture, whilst the consumption of kimonos the Japanese national costume, has declined. Production would have declined much faster, if the government, through the Japan Raw Silk and Sugar Stabilisation Agency (JRSSSA), had not established a cocoon and raw silk price support scheme and imposed restrictions on silk imports.

Whilst Japanese production is insufficient to meet domestic requirements, the JRSSSA has had to build up substantial raw silk stocks in recent years as political and economic factors have necessitated sizeable imports particularly from China and Korea. Japan's present silk production is to the tune of 5,000 tonnes only. During the last decade, the situation in the silk market has been a serious cause for concern both for the government and for these working in the industry. By mid-1984 consumption had started declining and 10,650 tonnes of raw silk were held in stock by the JRSSSA representing over a year's production. Within the industry and government there are conflicting views on the future policy. The sericulture farmers, who are influential in certain constituencies and can bring political pressure to bear, would like to see continued price support and stable production. Against them, however, are the silk weavers and traders who see the high support price as detrimental to their interest and would like to see an end to the monopolistic import system and cheaper silk prices, which they argue would facilitate an expansion of silk production and consumption and exports.

In 1983, the weavers organised a nationwide cutback of production to bring attention to their case, whilst the farmers arguments are supported by the Ministry of Trade and Industry (MITI). Both the MITI and the Ministry of Agriculture (MAFF) etc., have established study groups, the MAFF to look into the cocoon and raw silk stabilisation system and the MITI to look into the problems of silk consumption, distribution and international trade. Intense debate has been taking place in Japan about the future of the industry.

As Japan is in the forefront of rearing and reeling technology, a review in greater detail of its silk industry will be of interest to the existing and potential silk producers. Approximately 6,500 people are employed in the silk reeling mills of Japan, and there is a division of labour by gender in that cocoon processing is undertaken by men, whilst reeling and waste collection is undertaken by women.

Once produced, the raw silk is tested and classified at the conditioning houses, of which there are two national ones at Yokahama and Kobe, and several pre-fectural ones. There are several hundred dealers who transact business between the reelers and raw silk processing companies, with most transactions being undertaken at the Kobe and Yokahama Silk Exchanges. In 1980 raw silk processing was undertaken by approximately 2,000 throwsters, 16,000 weaving mills and 150 dyeing and printing concerns, but the recent difficulties in the silk industry have led to a decline in their number.[6]

According to estimates, the demand for raw silk in Japan in the year 2000 will be about 13,200 tonnes and domestic production of raw silk will be 7800 - 9000 tonnes leaving an import gap of 4200 - 5400 tonnes.[7]

Japan and Japanese markets will remain important to the international silk community. Japan has a long tradition both in silk production and consumption, and despite the decline in production, Japanese silk technology specially in silk processing will remain a valuable source of information to the developing countries aspiring for silk production and processing.

KOREA

Korea also has a long tradition as silk producing country, and today silk is viewed as a traditional industry that contributes both to farmers' income and to export earnings. When Korea was bifurcated into North and South Korea in 1945, most of the sericultural land remained with South Korea. North Korea did not look after silk cultivation, resulting in a total destruction of the silk industry under its field, whereas South Korea consolidated its sericulture activities and modernised them into a Japanese style industry. During the 1960s and 1970s silk production grew at a phenomenal rate mainly because of active governmental encouragement fall in Japanese production, and development of fears of a world shortage. Raw silk production peaked between 1975 and 1977 at over 5,000 tonnes per annum but declined quickly.

The silk industry of Korea largely catered to the Japanese market until Japan introduced import restrictions on silk products in 1972. After this the korean silk producers began to diversify and enter other markets such as those of Europe and America. In 1982, raw silk totalled 1,872 tonnes, from 12,547 tonnes of cocoons produced by 1,65,500 sericultural households cultivating 23,728 hectares of mulberry. A similar quantity was produced in 1983.

Owing to rapid industrialisation, it is becoming difficult in Korea to allocate land for mulberry cultivation and to find the manpower for the labour intensive sericulture processes and cocoon production. Resultantly there has been sharp decline in the silk production.[8] Raw silk production in Korea from 1965 to 1992 is given in Table 3.3.

Table - 3.3

RAW SILK PRODUCTION IN KOREA (1965 - 1989)

S. No.	Year	Raw Silk in Tonnes
1.	1965	1,876
2.	1970	3,026
3.	1975	4,445
4.	1980	3,319
5.	1985	1,503
6.	1988	1,343
7.	1989	1,200
8.	1992	1,380

Source : Korea Raw Silk Exporters Association, Seoul.

Table 3.3 shows that production of raw silk increased rapidly between 1965 and 1975 when it reached 5,545 tonnes. It declined steadily there after to 936 tonnes in 1989. The decline in domestic production has encouraged exporters of silk products to look for raw material supply from other parts of the world. China and Hong Kong have made largescale imports to keep the silk industry running in Korea. As in the case of Japan, silk is the basic material for the national costume, the 'Chima Jeokori'. The decline in demand for silk for the use of the national dress is being offset by the rise in domestic purchases of garments of silk designed in western styles. Domestic consumption of silk is, therefore, expected to expand. Moreover, it is customary to wear silk in marriages as in the case of India.

Recent price increases have made silk products from the Republic somewhat less competitive despite the rise in quality standards. It may be mentioned that the Korean Sericultural Association is actively employed in promoting the use of silk among local consumers. The association has special funds for various promotional activities in the country.

BANGLADESH

Sericulture in Bangladesh,which was part of India till about five decades ago, is as old as Indian sericulture. It has passed through different stages of rise and fall with the changes in the political, social and economic scene in the past 100 years. Raw silk production which was to the tune of 87 tonnes in 1876 was reduced to 0.05 tonnes in 1947. The sericulture industry in Bangladesh, after the country's liberation and independence in 1971, is making all efforts to attain its past glory.[9]

The Bangladesh Sericulture Board established in 1977 is responsible for the overall development of Sericulture industry in the country. The popular mulberry cultivation practices are bush and tree plantations. Owing to the growing demand for fertile, cultivable land for growing food and vegetable crops, bush plantation is becoming less popular. Unfavourable weather conditions, low price of cocoons, poor returns etc., discourage farmers to take up sericulture as a cash crop. Therefore, bush plantation is limited to traditional areas like Bholahat adjacent to Malda, the sericulture district of West Bengal in India.

The current emphasis in Bangladesh is to popularise mulberry tree plantation on fallow and non-cultivable land along the road side, railway lines, banks of canals and ponds and places around houses. 'Telea' is the local mulberry variety that is exploited to the maximum in both the bush and tree systems of cultivation. The total mulberry area were 3980 hectares in 1993, of which hardly 380 hectares were under bush plantation and tree plantation was in about 3500 hectares.[10]

At present the income through sericulture is not very attractive. Government and private organisations are in favour of developing sericulture as a subsistence crop, suitable to practice on fallow and uncultivable land, to engage the millions of rural population, especially the women, to generate extra income for many household workers, to employ thousands of artisans and weavers, with the ultimate motto of exporting finished silk goods to external market to earn the much needed foreign exchange.

THAILAND

Thai silk has acquired an international reputation but local production has failed to keep pace with the demand. Sericulture and silk weaving are traditionally practised in the economically poor north and north-east regions of the country and an estimated 0.5 million farmers undertake production with around 65,000 hectraes of mullberry under cultivation. The industry creates direct employment for about two million people and for another one million indirectly. Production is still done mainly by traditional methods handed down through families. Reeling and weaving is mostly by hand, with the weavers using either the traditional thrown shuttle looms or fly-shuttle looms numbering over 30,000 and 23,000 respectively. The resulting yarns have many knots and ties and these features help to give the Thai silk fabric its unique characteristic.[11]

Production of Thai raw silk, according to government statistics, is around 800 - 900 tonnes whilst cocoon production is around 8,000 -9,250 tonnes Shortage of warp yarn is a serious problem for Thailand. Thailand imports silk yarn ranging from 400 tonnes to 500 tonnes from China.

NEPAL

Sericulture in Nepal is still in the budding stage and despite Korean assistance, a proper infrastructure for sericulture development in terms of research, extension, breeding and processing, is still in the formative stage. There is a small domestic demand for silk from the weaving industry and despite the existence of a small reeling facility, the lack of local cocoon production has necessitated imports of raw silk.

SRI LANKA

Sericulture and silk weaving are new to Sri Lanka. But since the early 1970s efforts have been made to develop an integrated silk industry with financial assistance from several international organisations. In the late 1970s annual cocoon production reached 20 tonnes with Japan the major importer.

WESTERN COUNTRIES

In the earlier days the silk trade was confined to Asia though the west was very much fascinated by the silk fabric. But no one except China seemed to know anything about silkworm rearing. The silkworm was first mentioned in western literature by Aristotle. Later after the invasion of Asian countries by some of the European emperors like Alexander, paved the way for the spreading of silkworm rearing technique to Europe and elsewhere. Several countries, mostly in Eastern Europe, continue to produce silk but their contribution to the total world production is minimal and mainly dependent on continued State assistance.[12]

The sericulture industry spread through 'Italy' during the 12th and 13th centuries. For 300 years, Italy practically controlled the European silk technology. The advancement in silk processing in Italy is unmatched. The Italian silk industry is strongly export-oriented. It exports high quality silk goods which are appreciated world over. 'Milan and 'Como' are production and export centres for silk products.

Italy imports raw silk to the tune of 4500 tonnes and silk waste about 3500 tonnes. The spun silk industry in Italy is well established and mainly depends on China and erstwhile Soviet Union for its silk waste supply. India too used to export quite a substantial quantity of silk waste to Italy.

As one of world's fashion leader, Italy by all means will continue to expand its production of high quality fashionable silk goods. The United States continues to be the principal buyer of Italian Silk goods.

Italy is looking for an alternative source of raw silk supply. Its authorities are in touch with countries such as Vietnam and Zimbabwe, exploring the possibilities of raw silk supply.

From Italy sericulture slowly spread to France. During 18th Century France enjoyed a prosperous silk trade. 'Lyon' was the centre for sericulture industry. France established several spun silk mills. When sericulture was at its peak, at the end of 18th century, a mysterious disease broke out and wiped out the entire silkworms in France and later in the whole of Europe.

In modern times, European countries, instead of practising sericulture, prefer to import raw silk on a large scale from China and convert it into fabric and garments. Germany, Italy, France, Switzerland and U.K. are Europe's leading countries of silk. But unlike Italy and France, Germany does not have many processing units and hence imports finished products from other countries. As the silk weaving and processing activities are at its minimum in Germany, there exists a potential market in that country for the silk exporter. Germans prefer natural fabric to artificial ones. Among them, in terms of per capita, consumer spending is high on clothing and decoration. Perhaps these two may be the reasons for increased use of natural silk cloth in Germany.

Switzerland plays a significant role in the international silk trade. Quite large quantities of silk fabric and raw silk yarn are imported by Switzerland. It has a small silk processing unit at Zurich, fabricating fine silk products of its own choice.

Brazil is the fifth largest producer and second largest exporter of raw silk At present it is the only country outside the Asian continent which is steadily increasing the production of silk. Commercial silk production was started earlier in this century by Italian and Japanese immigrants.Silk production in Brazil has fluctuated widely and mainly on account of its reliance on export markets. After rapid expansion during the Second World War, production collapsed as the world markets became normalised. It was revived in the early 1960s, initially with exports to Europe, but later to Japan

Sericulture is carried out in the states of 'Sau Paulo' and 'Parane' in Brazil. More than 95 per cent of the country's silk is produced in these two states where the climate is temperate and the soil is well suited for mulberry cultivation.[13] The annual output of silk accounts for only three per cent of the world's silk production. However, silk production in the country is steadily increasing.

Table - 3.4 illustrates the supply and demand position of natural silk in Brazil. During 1980, the production of mulberry silk was 1,284 tonnes and it rose to 1,900 tonnes in 1989 and 2,300 tonnes in 1992. Accordingly there is a steady rise in exports as well as domestic consumption. The domestic consumption rose from 430 tonnes in 1980 to 550 tonnes in 1990.

Table - 3.4

SUPPLY AND DEMAND POSITION OF NATURAL SILK OF BRAZIL

(In tonnes)

S.No.	Year	Production	Exports	Domestic Consumption
1.	1980	1284	854	430
2.	1983	1347	1047	300
3.	1986	1633	1120	373
4.	1988	1740	1270	470
5.	1989	1900	1400	500
6.	1990	2200	1050	550

Source : T.D. Koshy - Silk Exports and Development, P.13.

Brazilian silk is of good quality and also attractive to foreign buyers. For years Brazilian raw silk has been finding entry into European markets as an alternative to Chinese silk. Brazil is forced to export more than two thirds of its silk production in the form of raw silk and silk yarn, as silk weaving and processing facilities are not available within the country. Silk in the country is progressing well both in quality and quantity. It has to be admitted that Brazil has broken the monopoly of Asian silk production.

CONCLUSION

Silk as a magical fibre has had great appeal to people all over the world for centuries. Today, it is being practiced in 58 countries of the world. Although the production of silk is rising every year it is not available in large quantities so that the share of silk in the total world textile fibre is only of the order of 0.17 per cent.

In the world at present sericulture has rapid growth mostly in the South East Asian Countries, which account for about 87 per cent of the world's raw silk production. China is recognisesd as the birth place of silk and the Chinese managed to keep the secret of silk to themselves for centuries. Subsequently, gradually other countries of the South East Asia and also some Western Countries too got to know it. All the same China maintains even toady its supremacy producing 63 per cent of world's total raw silk production.

Next to China, Japan was considered at one time to be the second home of silk, and it maintained for a long time its position as the second largest producer after China. But due to rapid industrialisation sericulture industry has declined and at present Japan's silk production is only of the order of six per cent and it ranks next to India (16 per cent).

Development of Sericulture in other South East Asian countries like Korea, Bangladesh, Nepal and Sri Lanka is very marginal and efforts are being made by the respective governments to popularise this industry as sericulture brings various benefits for the upliftment of rural economies.

Among the Western countries Italy, France, Germany, Switzerland and Brazil have some importance with regard to sericulture development. Brazil has the distinction of being the fifth largest producer and the second largest exporter of raw silk in the world. Brazilian silk is of good quality and attracts foreign buyers.

Note:-

1. T.D. Koshy, Silk Exports and Development, Ashish Publishing House, New Delhi, 1993, p.25.
2. CSB: Statistical Biennial, Silk in India, 1992, Bangalore, p.3.
3. T.D. Koshy, Silk Exports and Development, Op.cit., p.4.
4. Thnang Dahnan, Lin Shi Xian and Lilong, "Sericulture Production Strategies in the 21st Century", *Indian Silk*, Vol.33, No.8, p.5.
5. Ibid., p.22.
6. Ibid., pp.23-26.
7. T.D. Koshy, Silk Exports and Development, Op.cit., P.4.
8. Ibid., p.11
9. Dr. K.V. Benchaman, "Sericulture in Bangladesh", *Indian Silk*, Vol.32, No.6, 1993, p.17.
10. Ibid., P.19.
11. Peter Greenhaigh, The World Market for Silk, Op.cit., P.35.
12. Peter Greanhalgh, The World Market to Silk. Op.cit., P.35.
13. T.D. Koshy, Silk Exports and Development, Op.cit.,P.12.

CHAPTER 4
Progress of Sericulture in India

Attracted by the elegance of silk, princely rulers like Tippu Sultan of Mysore popularised and encouraged silk activity in India. Later on after Independence, the Government of India identified the silk industry as one of the employment-oriented cottage industries suitable for the development of rural India, and has been giving it a prominent place in the developmental plans. This has helped the industry to grow at a faster rate within the country and also spread to non-traditional areas.

India is now the second largest producer of silk next to China, and the only country producing all the four major commercial varieties of silk. Today, Indian sericulture provides an excellent model for other tropical countries in the production of mulberry raw silk.

EMERGENCE OF SERICULTURE IN INDIA - A HISTORICAL OVERVIEW

Sericulture in India is as old as its ancient culture. In India, some of the older Sanskrit works like The 'Mahabharata' describe 'Patta Keetas' (or silkworms), as having been brought by Chinese visitors as a present to King Yudhishtira. By the Mouryan period (fourth to second centuries B.C.) there was a trade in Chinese silks and a few centuries later Indian silks seemed to have been much in demand in the ancient Roman Empire.[1]

Historians consider it probable that mulberry sericulture was first introduced in India as late as the fourteenth and fifteenth centuries. Though wild silks (Tasar, Eri and Muga) were produced in India for a long time, mulberry sericulture is thought to have been introduced in this country fairly late and through a long and circuitous route.[2] The first firm evidence of sericulture in Kashmir is contained in the 'Tarikh-i-Rashidi', completed in 1547.

The earliest evidence of the production of mulberry silk comes from the Mughal period, during which the Industry had a prosperous time. An estimate given by Jean Baptist Tavernier of the middle of the seventeenth century suggests that the annual supply of silk at Kasimbazar (North

Bengal), then the principal trading centre, ranged from 2.4 to 3.1 million pounds (1100-1400 tonnes) by weight. In the seventeenth century there was an extensive overland and overseas commerce in silk along with other traded commodities like cotton, indigo and opium. At this time, Gujarat was the main producing centre for finished silk, while small quantities were produced in Kashmir, Bengal, Varanasi, Lahore and Fathepur.[3]

By the latter half of the seventeenth century, Bengal replaced China as supplier of raw silk for the Indian looms. Silk factories were established at Kasimbazar and the Dutch found a market in Japan for the silk produced in India. This was regarded as "Bengal Silk Trade". Britishers started the export of Indian silk to Europe. Thus Indian silk products became world famous.

In the eighteenth century, filature reeling units were established at Rangapur and Mushirabad. Filature reeling was introduced by the East India Company in 1769 and Italian craftsmen were employed to train Indian reelers. During the eighteenth and nineteenth centuries changing levels of sericulture were dictated by changes in the silk market. The Bengal silk industry suffered a serious set back during the revolutionary wars. Between 1792 and 1811 the silk trade stagnated and improved after the imposition of the continental blockade which stopped silk imports into London from Italy. By the middle of the nineteenth century, China had replaced Bengal as the major supplier of Europe's silk needs. Competition on both price and quality with China, led to a decline in activity by the 1870s, and as a result there was a cessation of raw silk exports from India by the First World War.[4]

The Second World War, however, provided a great impetus to the Indian sericulture industry. With the stoppage of supplies of raw silk, an item essential for the production of parachute fabric and its components, from Japan, China and the European countries, the allies had to look to India for meeting the demand for this strategic raw material. The Government initiated the filature expansion scheme and also introduced other measures for stepping up the production of raw silk. The area under mulberry increased from 17,628 hectares in 1937-38 to 45,581 hectares by the end of 1945. Simultaneously raw silk production also increased from 0.69 tonnes (1937-38) to 0.95 tonnes (1945)[5].

PROGRESS OF SERICULTURE IN INDIA UNDER FIVE YEAR PLANS

With the advent of economic planning in the country, the growth of sericulture industry has been accelerated through appropriate programmes. The following pages trace the growth achieved by this industry during the differenent plan periods. The Plan allocation and expenditure on sericulture development in India is presented in Table - 4.1.

Table 4.1 shows how the allocation of funds for sericulture has gone up since 1951-52 to 1987-88. During the First Five Year Plan, the amount allotted for sericulture was only Rs.60.22 lakhs, of which an expenditure of Rs.33.64 lakhs (56 per cent), was incurred. Since then the funds from the Government of India for sericulture has been increasing. During the Seventh Five Year Plan Rs.31,078.00 lakhs were allotted. By the end of 1987-88 Rs.15,892.24 lakhs were spent on sericulture development through different programmes.

The amount approved for sericulture development by the Planning Commission for both Central and State Government schemes and the utilization of the funds under Central and State schemes since the First Five Year Plan are shown in Table 4.2.

Table - 4.1

PLAN ALLOCATIONS AND EXPENDITURE ON SERICULTURE IN INDIA

(Rs. in lakhs)

Sl. No.	Particulars	Allocation	Expenditure	Percentage of Utilisation
1.	First Five Year Plan	60.22	33.64	56
2.	Second Five Year Plan	414.37	250.62	60
3.	Third Five Year Plan	702.00	411.00	59
4.	Annual plans (1966-69)	552.51	281.73	51
5.	Fourth Five Year Plan	969.00	674.99	70
6.	Transitional period (1978-80)	2,554.00	1,695.00	66
7.	Sixth Five Year Plan	17,637.00	12,638.00	72
8.	Seventh Five Year Plan	31,078.00	29,404.63	95

Source : CSB: Silk in India - Statistical Biennial, Central Silk Board, Bangalore, 1988.

Table - 4.2

PERCENTAGE OF UTILISATION OF FUNDS UNDER BOTH CENTRAL AND STATE SCHEMES ON SERICULTURE

(Rs. in lakhs)

PARTICULARS	CENTRAL SCHEMES			STATE SCHEMES		
	Allocation	Expenditure	Percentage of utilisation	Allocation	Expenditure	Percentage of utilisation
First Five Year Plan	14.25	11.95	84	45.97	21.69	47
Second Five Year Plan	35.17	26.13	74	379.20	224.49	59
Third Five Year Plan	150.00	72.00	48	552.00	339.00	61
Transitional Period (1966-69)	70.23	62.75	89	482.28	218.98	45
Fourth Five year Plan	130.00	81.61	63	839.00	593.38	71
Fifth Five Year Plan	861.00	586.17	68	1,693.00	1,109.06	66
Transitional Period (1978-80)	1,320.00	1,105.46	84	1,589.71	1,259.26	79
Sixth Five Year Plan	4,000.00	3715.00	93	13,637.00	8,923.00	65
Seventh Five Year Plan	7,000.00	8,816.00*	-	24,078.00	20,588.63	86

Source: Silk in India, Statistical Biennial, Central Silk Board, Bangalore 1992.
Note: * Includes an expenditure of 240.00 lakhs under NSP.

Table 4.2 reveals that the amount allocated under Central schemes for sericulture development was, by and large, well utilised. But the utilisation under State schemes was slow in the beginning, between 1951-52 and 1968-69. But since 1970, the State schemes show an immense improvement both in allocation and utilization of funds for sericulture development. A detailed plan-wise description of allocation of funds and targets achieved is presented in the following pages.

FIRST PLAN

Sericulture did not find a separate place in the First Five Year Plan, and was included under 'Other Village Industries'. Central assistance was made available to the States by the Board as grants and loans for implementation of specific approved schemes. During the plan period, a total amount of Rs.45.97 lakhs was made available to the States, of which Rs.21.69 lakh was utilised. The percentage utilisation of the outlay made available was low due to paucity of trained personnel and delay in constructional activities. The developmental schemes initiated during the period were designed to consolidate the industry and provide the necessary organisational base. In all, about 130 schemes were implemented in the different states. Under the Central Sector, there was a provision of Rs.14.25 lakhs of which Rs.11.95 lakhs were utilised. Out - put of raw silk rose from 8.94 lakh Kg. at the beginning of the plan to 14.86 lakh kg. at the end, accounting for a 66 per cent increase.[6]

SECOND PLAN

Out of an allocation of Rs.379.25 lakhs during Second Plan for the implementation of 339 schemes, about Rs.224.49 lakhs (nearly 60 per cent) were utilised by the states. In the central sector, the outlay was Rs.35.17 lakhs and the expenditure was Rs.26.13 lakhs. Production of raw silk increased from 14.86 lakh Kg. to 15.13 lakh Kg during the plan period.

Emphasis was laid on the development of seed organisation and improvement of silk reeling. Among the major achievements mention should be made of the passing of legislation in Karnataka to prevent the use of unexamined seed for improving the quality and output of raw silk, establishment of the Central Silkworm Seed Station at Pampore (Srinagar), and the setting up of the All India Sericulture Training Institute at Mysore (since merged with the Central Sericultural Research and Training Institute) by the Central Silk Board.

The Government of India, on the advice of the Board, canalised the imports of raw silk and its distribution to regulate the imports and stabilise the domestic markets. In the export sector, for the first time, an organised effort was made to promote exports with the introduction of export promotion schemes, administration of which was entrusted to the Board. Accordingly, the Board established five inspection centres in September 1958.[7]

THIRD PLAN

The Plan period assumes an importance in the development of sericulture in India since it was only during this period that sericulture was assigned a separate place within the overall budget head of 'Village' and 'Small Scale Industries'. Export promotional efforts were further enlarged both in their nature and content.

An outlay of Rs.7.02 crore was approved for the development of the industry during the Third Plan period of which Rs.5.52 crore related to State Schemes and Rs.1.50 crore for Central Schemes to be implemented through the Central Silk Board. The total expenditure incurred by the State was Rs. 3.39 crore accounting for 61.4 per cent of the allocation. An expenditure of Rs. 0.42 crore was incurred in respect of the Central schemes.

During the plan period, the seed organisation was further strengthened in the States of Jammu & Kashmir and Karnataka. Organisation of research received particular attention and the Board organised the Central Sericultural Research and Training Institute at Mysore (1961). Modernisation of reeling with a view to improving the quality of raw silk made considerable progress. On the export front, quality control inspection and standards were introduced in January, 1965. 'The Central Silk Trade Mark', scheme for quality was also introduced at the same time for implementation on a voluntary basis to prevent sales of art silk fabrics as pure silk fabrics by unscrupulous traders.

ANNUAL PLANS (1966-69)

. During this period, under sericulture, the allocation for State was Rs. 482.28 lakhs and the utilisation was Rs. 218.98 lakhs. The allocation for Central schemes was Rs. 70.23 lakhs and the utilisation was Rs. 62.75 lakhs. Output of raw silk rose from 20.65 lakh kgs. at the end of third plan to 23.20 lakh kgs. at the end of the period 1968-69 accounting for 12.3 per cent increase.

FOURTH PLAN

During this plan period, the allocations to the States and Central projects were Rs. 839 lakhs and Rs. 130 lakhs respectively. The corresponding utilisation was Rs. 593.38 lakhs and Rs. 81.61 lakhs. The objective of the Board during IV Plan was the attainment of self-sufficiency with regard to the country's production of raw silk through increased productivity, reduced cost of production through rationalisation of the production techniques and creating additional employment opportunities to about 4.00 lakh persons.

Under the Central sector, the research and seed organisations were further strengthened. Regional Research stations were set up at Majra (U.P.) for mulberry. The most important achievements during the plan period are introduction of Oak tasar in the sub-Himalayan regions of the States of Jammu & Kashmir, Himachal Pradesh, Uttar Pradesh and introduction of bivoltine rearing in the traditional multivoltine areas of Karnataka, Tamil Nadu and West Bengal. The Central Silk Board established the Raw Material Bank at Chaibasa in Bihar in 1972-73, in order to impart a measure of stability in the tasar cocoon market conditions.

By the end of the plan period, production of raw silk reached 28.94 lakh kgs. and export-earnings reached Rs. 14.46 Crores.

FIFTH PLAN

The Planning Commission approved an outlay of Rs.25.54 crores during the Fifth Five Year Plan period. Out of this, a sum of Rs. 1109.06 lakhs was spent on the State schemes and Rs. 586.17 lakhs on the Central schemes. The Central Silk Board introduced price stabilisation schemes for mulberry silk during the year 1977-78 for ensuring a fair economic return to the primary producers and availability of raw silk to silk hand-loom weavers, manufacturers and exporters at a fair and steady price. By the end of the Plan period production of raw silk reached the level of 37.11 lakh kg. and export earnings touched the level of Rs. 33.06 crores.

ANNUAL PERIODS (1978-80)

An allocation of Rs. 1248.05 lakhs was made available for the year 1978-79. However, the total utilisation amounted to Rs. 985.39 lakhs. (Rs. 510.36 lakhs were utilised on States schemes and Rs. 475.03 lakhs on Central projects which included Intensive Sericulture Development Projects sponsored and administered by the Boards in the States.) During the year, production of raw silk increased to 41.77 lakh kgs. from 37.11 lakh kgs of the preceding year . On the export front, the achievement was Rs. 43.67 crores as against the target of Rs. 37 crores.

The Planning Commission approved an allocation of Rs. 1661.66 lakhs for sericulture development for the next year 1979-80 of which Rs. 861.66 lakhs were for the States and Rs. 800 lakhs for the Central projects. However, out of the outlay for the Central projects, Rs. 400 lakhs were transferred to the States for the continuance of intensive sericulture developmental schemes. Against this, the expenditure was Rs. 1379.33 lakhs, Rs. 748.90 lakhs under the State schemes and Rs. 630.43 lakhs on the Central projects. The production of raw silk rose to 48.05 lakh kgs. from 41.77 lakh kgs. during 1978-79. On the export front, the achievement amounted to Rs. 48.83 crores as against the target of Rs. 50 crores.

SIXTH PLAN

The Planning Commission approved an allocation of Rs. 167.37 crores for the Sixth Plan period. However, the year to year allocation approved under the Central projects during the Plan period amounted to Rs. 40 crores. Thus the allocation during the Plan period as a whole amounted to Rs. 176.37 crores. Out of this, a sum of Rs. 89.23 crores was spent on the State schemes and Rs. 37.15 crores on the Central projects. By the end of the plan period, production of raw silk reached the level of 76.73 lakh kgs. and silk export earnings rose to Rs. 129.05 crores.

The National Silkworm Seed Project came into existence in 1981 so as to expand the activities of systematic seed organisation and to develop technologies for bivoltine sericulture propagation in the country.

SEVENTH PLAN

The Planning Commission, approved an allocation of Rs. 310.78 crores for the Seventh Plan period. By the end of the Plan period, production of raw silk reached the level of 12,016 tonnes and silk export earnings amounted to Rs.400.61 crores.

Intensive Sericulture Development Projects for mulberry were taken up for implementation in Orissa and West Bengal at a cost of Rs. 427 lakhs and Rs. 967 lakhs respectively. The amounts were intended for expansion of area under improved variety of mulberry, setting up of sound infrastructure for quality seed production, supply and marketing, etc.

During 1989-90, the closing year of the Seventh Plan period, The Central Silk Board with the financial assistance of World Bank and The Swiss Development Co-operation (SDC), for the development of mulberry sericulture in the country, initiated 'The National Sericulture Project' (NSP) at a total cost of Rs. 555.3 crores. The project envisaged to bring an additional area of 57,600 hectares of land under mulberry cultivation in order to produce additional 6,072 tonnes of mulberry raw silk by the end of the five-year project period.

TRANSITIONAL PERIOD (1990-92)

The Planning Commission, approved an allocation of Rs. 161.20 crores for the development of sericulture both under the State and Central sectors, for the year 1990-91. During the year production of raw silk increased to 12,665 tonnes from 12,061 tonnes in 1989-90. On the export front, the achievement was Rs. 440.53 crores as against the previous year's achievement of Rs. 400.61 crores.

For the year 1991-92 the allocation approved under both the State and Central sectors was Rs. 211.88 crores and the physical targets suggested were 14,060 tonnes of raw silk production and export earnings of Rs. 600.00 crores.

EIGHTH FIVE YEAR PLAN

The Planning Commission constituted a sub-group on sericulture for formulation of sericulture development programmes for the Eighth Plan. The final outlay proposed for the development of sericulture during the Plan period (1992-97) both under the Central and State Sectors is Rs. 860.76 crores[8] Table 4.3 gives the details.

Table - 4.3

PROPOSALS OF EIGHTH FIVE YEAR PLAN (1992-97)

Sl. No.	Item	Allocation (Rs. in crores)	Targets (as specified)
I	FINANCIAL		
1.	States Plan Programmes	583.55	-
2.	Central Sector Provision	277.21	-
3.	Total Allocation	860.76	
II	PHYSICAL		
4.	Mulberry area expansion		
	a. Irrigated	-	3,00,000 ha
	b. Rainfed	-	1,25,000 ha
5.	Raw Silk Production		
	a. Mulberry	-	
	1. Multi x Bivoltine	-	18,000 tonnes
	2. Bivoltine	-	2,000 tonnes
	b. Non-mulberry	-	
	1. Tasar	-	650 tonnes
	2. Eri	-	650 tonnes
	3. Muga	-	100 tonnes
6.	Employment generation	-	65 lakh persons (cumulative)
7.	Export	-	Rs. 0.74 crores

Source : Silkman's companion, C.S.B., Bangalore

TRENDS IN MULBERRY SERICULTURE

The area under mulberry cultivation in India has been gradually increasing for the last two decades. In 1980-81 the area under mulberry crop was only 1,70,000 hectares and it increased to 3,41,200 hectares in 1993-94. The figures of reeling cocoons and raw silk production also have been showing a continuous upward trend as may be seen in Table - 4.4.

Table - 4.4

PROGRESS OF MULBERRY SILK IN INDIA

(Production in tonnes)

Year	Area under mulberry (in thousand ha.)	Reeling cocoons	Raw silk
1980-8	170.00	58,208	4,593
1981-82	179.95	55,210	4,801
1982-83	196.85	66,811	5,214
1983-84	206.91	71,276	5,681
1984-85	214.84	74,875	6,895
1985-86	217.84	76,717	7,029
1986-87	229.76	81,573	7,905
1987-88	241.60	86,528	8,455
1988-89	268.06	96,471	9,683
1989-90	294.24	1,10,433	10,905
1990-91	313.11	1,16,672	11,487
1991-92	331.24	1,07,153	10,658
1992-93	342.16	1,29,685	13,000
1993-94	341.20	1,23,157	13,392

Source : CSB: Silk Man's Companion, 1994, Bangalore, 1994.

Table 4.4 presents a clear picture of the trends in the progsress of mulberry silk in India from 1980-81 to 1993-94. There is a steep rise in the area under mulberry, reeling cocoons and raw silk production (Graphs 4.1, 4.2 and 4.3). The growth rates are statistically analysed and presented in the Table 4.5

From Table 4.5 it is clear that the area under mulberry cultivation in India rose at the rate of 5.88 per cent per year between 1980-81 and 1993-94. The multiple R^2 0.982 is significant at 1 per cent level indicating that the growth rate is significant.

Table 4.5

GROWTH RATES OF AREA UNDER MULBERRY, REELING COCOONS AND RAW SILK

Category	Constant	Regression coefficient	Crowth Rate	R^2	t- values
Area under Mulberry	5.0792	0.057	5.881	0.982	25.755
Reeling Cocoons	3.9738	0.065	6.706	0.961	17.170
Raw Silk	1.4327	0.087	9.089	0.978	23.185

Source : Computed from the data drawn from Table 4.4

The growth rates of reeling cocoons and raw silk were 6.706 and9.089 respectively. The coefficients of determination (R^2) for reeling cocoons and raw silk during this period were 0.982 and 0.978 respectively. The correlation coefficients for all three categories are significant at 1 per cent level and the high correlation coefficient shows that there was a steady growth during the period under consideration.

Out of the 6.29 lakh villages in India Sericulture is being practised in about 59,528 villages employing 60.30 lakh persons. Table 4.6 shows the trends in employment in sericulture since 1977-78.

The vital role played by sericulture in employment generation can be seen from Table 4.6. During 1977-78, the persons employed in sericulture were only 38.06 lakhs. In a decade their number rose to 60.30 lakh persons. The rapid increase of persons engaged in sericulture proves that sericulture is a labour intensive industry. Graph 4.4 shows the trends in employment in the sericulture sector in India from 1977-78 to 1988-89. The Table 4.7 gives the statistical analysis of the trends in employment in sericulture.

It is evident from Table 4.7 that the employment rate in sericulture in India was increasing at the rate of 4.965 per cent per year from 1977-78 to 1988-89. The coefficient of determination ($R^2 = 0.937$) is significant at 1 per cent level indicating that the growth rate is significant.

The sericulture industry has now established itself as an export oriented sector of the country's economy with an annual foreign exchange earnings of about Rs. 440.53 crores in 1990-91.[9] The main items of export of the silk industry are silk fabrics, readymade garments, carpets, and silk waste. Exports of silk goods are permitted freely but the silk waste export is regulated keeping in view the requirements of the indigenous spun silk industry. In addition to the traditional markets in the Asian and African

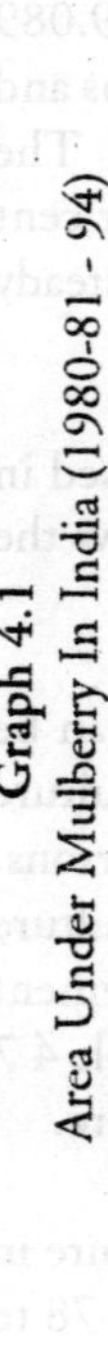

Graph 4.1
Area Under Mulberry In India (1980-81- 94)

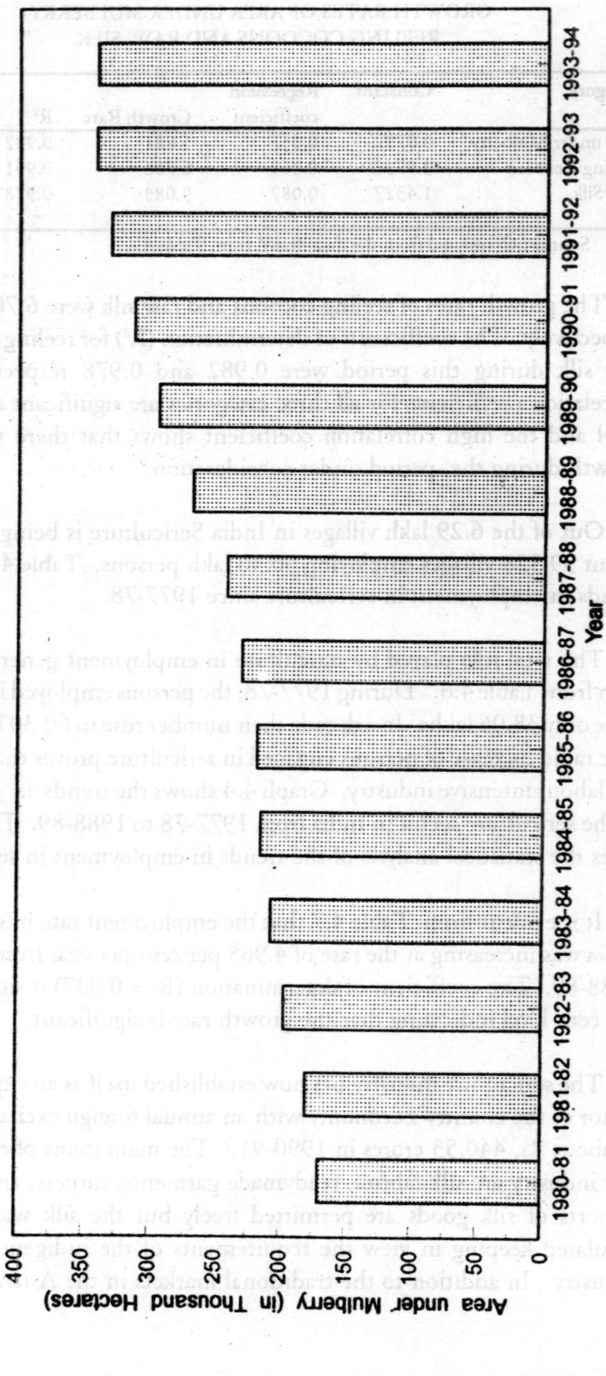

Graph 4.2

Production of Reeling Cocoons in India (1980-81 - 1993-94)

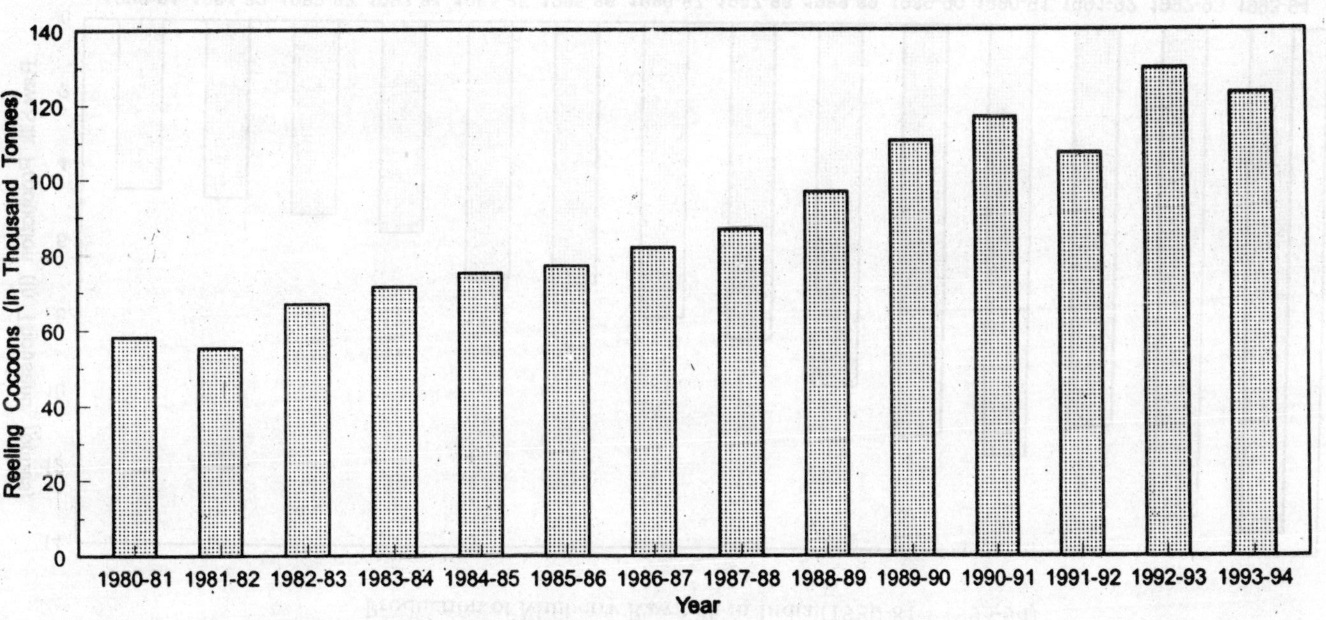

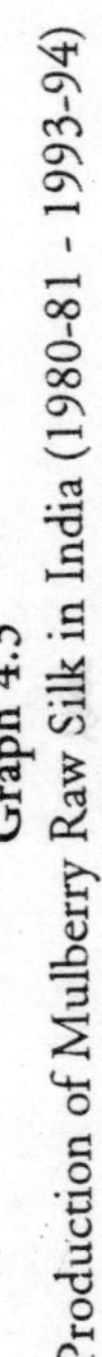

Graph 4.3

Production of Mulberry Raw Silk in India (1980-81 - 1993-94)

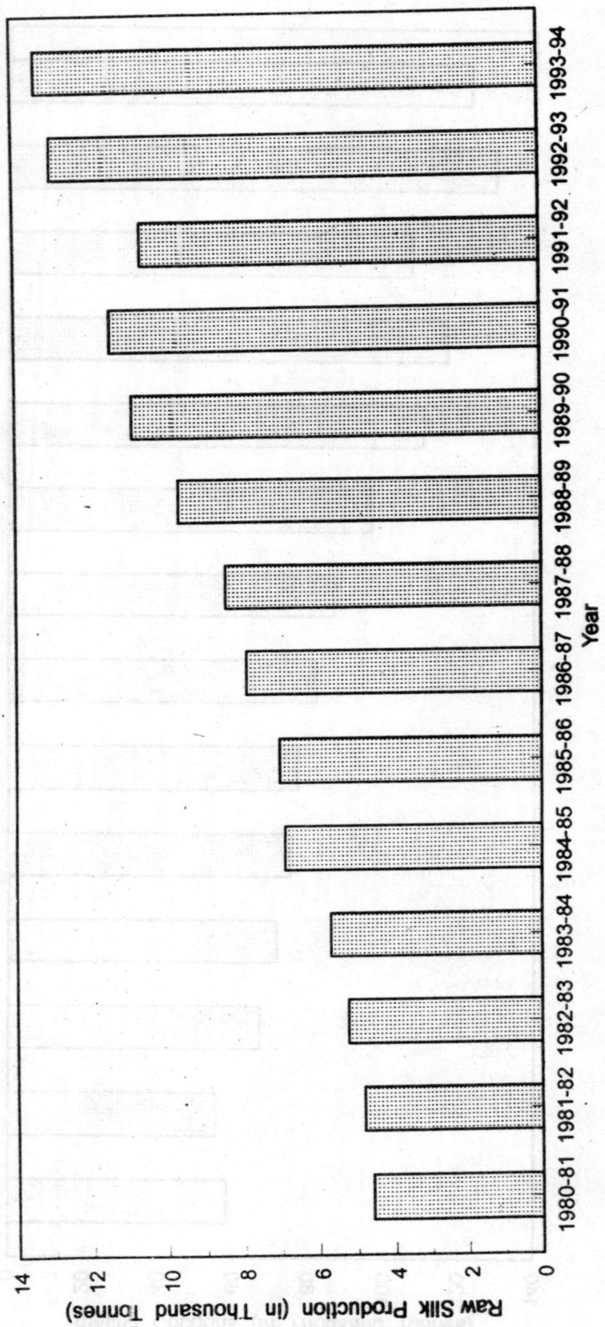

countries, Indian silk goods have found a good new market in the non-traditional countries such as the U.S.A., Canada, West Europe, Japan, Australia and Newzealand. The details of the export earnings of the Indian Silk Industry are shown in Table - 4.8

Table - 4.6

TRENDS IN EMPLOYMENT IN SERICULTURE IN INDIA (1978-89)

Year	Persons employed (in lakh numbers)
1977-78	38.06
1978-79	38.30
1979-80	40.00
1980-81	42.30
1981-82	44.55
1982-83	46.00
1983-84	50.50
1984-85	51.52
1985-86	53.64
1986-87	55.00
1987-88	57.65
1988-89	60.30

Source : CSB, Statistical Biennial, 1988, Bangalore.

Table 4.7

GROWTH RATE IN EMPLOYMENT GENERATION THROUGH SERICULTURE

Constant	Regression Coefficient	Growth Rate	R^2	t - value
	3.5615	0.048	4.965	0.937 12.139

Source : Computed from the data drawn from Table 4.6.

It is evident from Table 4.8 that the export earnings of sericulture showed a rapid increase during the last decade. During 1980-81 the value of silk goods exported was Rs. 52.95 crores and it rose to Rs. 435.95 crores in 1990-91. The total value of exports, both silk goods and silk waste, rose from Rs. 53.12 crores in 1980-81 to Rs. 440.53 crores in 1990-91. The growth rates are statisticaly tested and the results are presented in Table 4.9.

It is clear from the Table 4.9 that the coefficients of the value of silk goods and the total value of silk goods and silk waste are significant at 1 per cent level indicating that the growth of foreign exchange earnings on silk goods and on the total are significant and the growth rates are 24.550 and 24.176 respectively. In the case of silk waste the correlation coefficient is not significant implying that the growth rate is not significant even if it is 20.623.

Table 4.8

EXPORT OF SILK GOODS AND FOREIGN EXCHANGE EARNINGS

Foreign Exchange Earnings

Year	Silk goods		Silk Waste		Total Value
	Quantity (lakh Sq. m.)	Value (Rs. incrores)	Quantity (tonnes)	Value (Rs. incrores)	(Rs. in crores)
1980-81	125.82	52.95	133	0.17	53.12
1981-82	142.99	68.90	393	0.83	69.73
1982-83	140.90	79.31	1035	3.54	82.85
1983-84	147.75	96.30	2882	15.37	111.67
1984-85	170.70	125.33	291	3.72	129.05
1985-86	194.15	159.21	69	0.61	159.82
1986-87	243.53	200.01	489	1.48	201.49
1987-88	326.56	251.79	788	3.17	254.96
1988-89	361.42	327.92	604	2.62	330.54
1989-90	358.34	392.48	808	8.13	400.61
1990-91	324.54	435.95	384	4.59	440.53

Source: Silkman's Companion 1992, Central Silk Board, Bangalore, P.13.

Table - 4.9

GROWTH RATES OF VALUE OF SILK GOODS, VALUE OF SILK WASTE AND TOTAL VALUE

Category	Constant	Regression coefficient	Growth Rate	R2	t-value
Value of Silk goods	3.7477	0.220	24.550	0.996	48.177
Value of silk waste	-0.3081	0.187	20.623	0.242	1.694
Total Vlaue	3.7929	0.217	24.176	0.996	49.955

Source : Computed from the data drawn Table 4.8.

PROMOTIONAL AGENCIES AND SERICULTURE DEVELOPMENT

The role of sericulture in engaging the available abundant manpower in gainful employment is very significant. But in India whatever the economic activity may be, it can not make a head way and flourish without proper support from the Government. Moreover, in the years immediately following independence, mulberry sericulture was characterised by low productivity in terms of leaf yield as well as the quality of cocoons. The condition of non-mulberry sericulture was even worse, as it was confined to the hilly areas. The Government of India had set up the Silk Development Directorate in 1945. A Silk Panel was also established to suggest measures for sericulture development. This Silk Panel suggested a fifteen-year perspective plan for the development of sericulture and recommended the establishment of Central Silk Board for ensuring the co-ordinated development of the industry under Central control.

CENTRAL SILK BOARD (CSB)

The Central Silk Board was constituted by an Act of Parliament on 9th April 1949. Planned development of sericulture in India was taken up only after its establishment.

The Board has 36 members including the Chairman, Vice-Chairman, member-Secretary, representatives of the Lok Sabha and the Rajya Sabha, nominees of the Central and State Governments and representatives from among rearers, reelers, the trade and industry. Their term of office is three years. The Central Silk Board has been entrusted with the task of overall development of the silk industry.

The main functions of the Central Silk Board are: Promotion and development of silk industry by such measures as it thinks fit; undertaking, assisting and encouraging scientific technological and economic research relating to silk; developing and distributing healthy silk worm seeds; devising means for improved methods of mulberry cultivation, silkworm rearing, silk reeling and spinning; initiating measures of standardisation and quality control of silk and silk products; rationalisation of marketing and stabilisation of prices of silk cocoons and raw silk products for export; collection of statistics; preparing and furnishing relevant reports to the silk industry and to the Central Government; and advising the Central Government on all matters relating to the development of silk industry including the import of raw silk and export of silk products.[10]

The Central Silk Board has its head quarters in Bangalore. The offices of Chairman and Member-Secretary are located at the Head Office. It has established number of facilities all over the country to carry out Research

Graph 4.4

Trends in Employment in Sericulture in India (1977-78 - 1988-89)

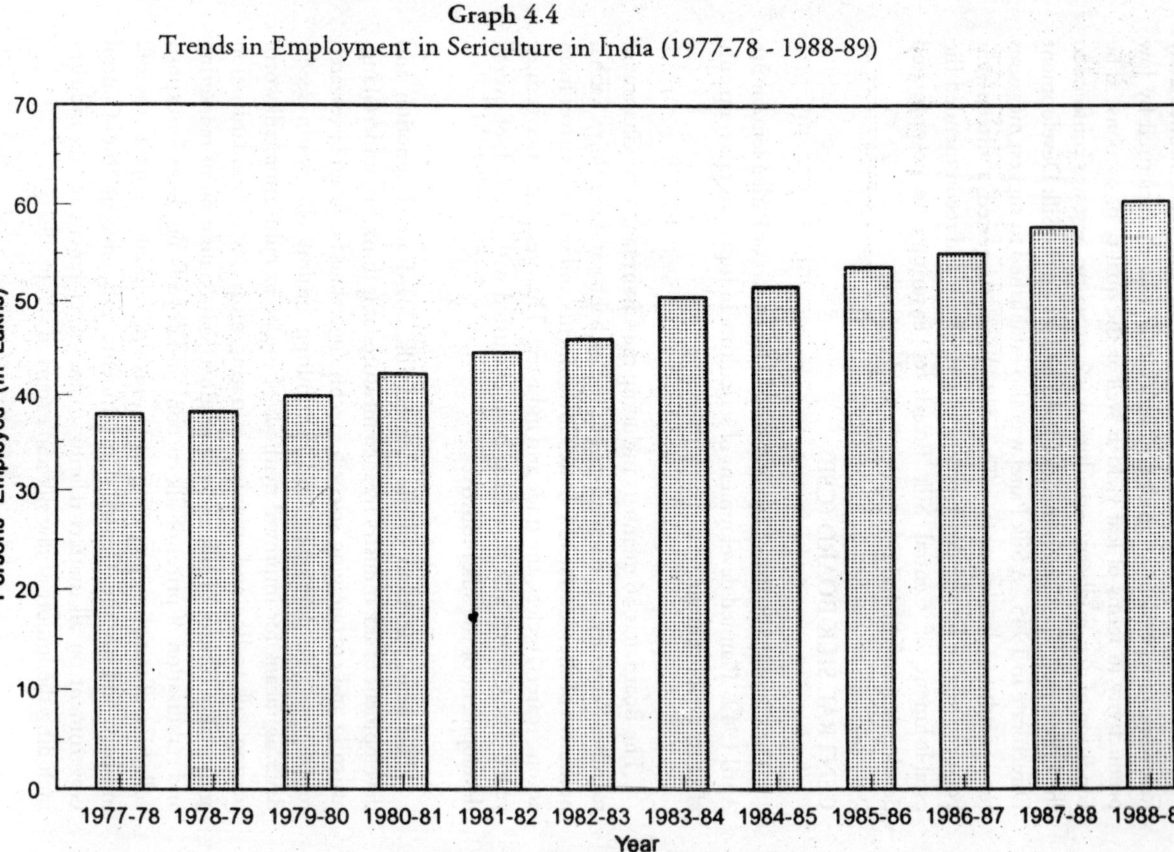

and Development in all aspects of the sericulture industry.

CENTRAL SERICULTURAL RESEARCH AND TRAINING INSTITUTE (CSRTI)

To develop tropical oriented research the Central Sericultural Research and Training Institute was established in Mysore in 1962. During the last 25 years the research institute has been able to evolve several high yielding mulberry varieties, utilising the local and exotic races. Mention may be made of S_{54} $S_{30,}$ S_{36} and S_{41} although Kanva - 2, one of the mulberry varieties evolved is not very popular with the sericulturists in South India.[11] The package of practices prescribed by this Institute after sustained efforts have helped in doubling the per unit mulberry leaf yield.

The new technology of silkworm rearing suitable to tropical conditions was developed in the seventies by the Institute. This helped in stabilising the cocoon crops, spread of sericulture industry in non-traditional areas and introduction of high-yielding silkworm races. A medicine called Resham Keet Oushadh has been developed and distributed extensively to control grasserie and muscardine diseases infecting the silkworm. To prevent the uzyfly infection, a product called Uzicide has been developed. These two are very popular among sericulturists.

NATIONAL SERICULTURE PROJECT WITH EXTERNAL ASSISTANCE (NSP)

The biggest impetus in recent times to the development of Indian sericulture was provided by the commencement of the five-year, Rs. 555 crore. National Sericulture Project (NSP) conceived by the Central Silk Board for the development of mulberry sericulture, and assisted by the World Bank and the Swiss Development Cooperation. The project has been under implementation from 1989 onwards in 17 major states in India. In the traditional sericulture states of Karnataka, Andhra Pradesh, West Bengal and Jammu & Kashmir, the project is implemented by their respective Departments of Sericulture. The Central Silk Board implements 12 pilot project in the states of Assam, Bihar, Gujarat, Kerala, Haryana, Himachal Pradesh, Punjab, Uttar Pradesh, Madhya Pradesh, Orissa and Rajasthan. The Board also executes several components such as improvement of Research and Development support, creation of additional facilities for training, silkworm seed production, germ plasm maintenance, extension etc., in the traditional states.

The objectives of the NSP as visualised by the board are: (1) introduction of sericulture in the other states, and expansion of area under mulberry by 58,000 hectares by the end of 1994, (2)increasing raw silk production in the country by an additional 6,000 tonnes by 1994, including 1,000 tonnes of superior bivoltine silk, (3) improving the quality and productivity of Indian silk by introducing specific measures, (4) generating employment opportunities in rural areas for an additional one million people, the majority of whom belonging to the economically weaker sections, (5) increasing the quantum of Indian Silk Product Exports by an additional Rs. 5,700 millions by 1994, (6) strengthening the infrastructure for research extension, seed production, silk processing, implementation of measures for quality control in the silk production system, and development of market support for cocoons as well as raw silk, (7) providing financial support to rearers, reelers, twisters and seed producers through commercial banks under the component of Working Capital Assistance to Reelers, (8) promoting increased participation of private sectors in silkworm seed production, supply of Chawkie reeling equipment, and (9)strengethening socially desirable features such as development of women, promotion of smokeless chullhas in reeling establishments, improvement of water conservation and rainfed technology, and involvement of Non-Governmental Organisations for promotion of sericulture.

Table 4.10 indicates the plan of financing the National Sericulture Project:

From Table 4.10 it is clear that an amount of Rs. 5553 millions was allocated through different financial agencies for sericulture development. The emphasis was laid on the I.B.R.D/I.D.A. loans and Institutional Credit which together constitute 81 per cent. The main thrust of the NSP in the traditional states was directed towards consolidating the progress already achieved by strengthening the existing infrastructure and establishing new facilities for further expansion. Table 4.11 gives the details of targets for mulberry acerage, cocoon and raw silk production.

Table 4.10
FINANCING PLAN UNDER NSP

Funding Agency	Rs. in Millions	Percentage to Total
I.B.R.D./I.D.A. Loan	2,832	51
SDC grant	400	7
GOI/State Investment	657	12
Institutional Credit	1,664	30
Total	5,553	100

Source: Silkman's companion 1992, Central Silk Board, 1992.

The NSP in the pilot states is designed to provide comprehensive packages with relevant backward and forward linkages, right from basic seed production to the production of yarn. A component of assistance to reelers is also built into the project.

The NSP has been showing particular attention to the role of women in sericulture and has worked out programmes towards that end. Sericulture is an agro-based industry with a natural integration of interdependent rural, semi-urban and urban based activities such as cocoon production, cocoon reeling, yarn twisting weaving and finally trading. All these activties result in employment at the rate of about 13 work years per hectare of mulberry, out of which an estimated 60 per cent is set apart for women. This enhanced role of women in sericulture is significant since it is in contrast to the other agricultural or agro-based professions where the participation of women is comparatively less.

This higher proportion of women's participation in sericulture has so far been a natural and self-regulated phenomenon. Appropriate interventions and assistance can add new dimensions to this endeavour and options available may further be expanded for improving the status of women in family enterprises. As part of such interventions, the CSB has built into the National Sericulture Project, a special component of Assistance to women and the NGOs.[12]

The sericulture industry by its nature is a blend of so many activities both on-farm and non-farm and requires support systems like, training, extension, infrastructural support, information support, technology, credit availability, timely raw material supply, market position and support of skilled persons etc. Though the Government has been providing all these supports in considerable measure, the government system itself has certain limitations. For example, there are problems of motivation, shouldering additional responsibilities which affect directly or indirectly the process of development, which lie outside its purview.

In this context, the Non-Governmental Organisations (NGOs), can and do play a vital role by raising the level of technology, offering superior technology and training, motivating and organising the farmers for such activities as technology absorption, narrowing the gap between the laboratory and land, etc.

Financial assistance is being provided to the bona fide NGOs and women groups to take up sericulture. The component of assistance to the

Table - 4.11

TARGET FOR MULBERRY ACERAGE, COCOON AND RAW SILK PRODUCTION
(Unit: acerage: ha; Cocoon & Raw silk production: tonnes)

State	Status at Commencement	NSP Target incremental	Year-wise targets				
			1989-90	1990-91	1991-92	1992-93	1993-94
Karnataka							
Mulberry Plantation	140456	15000	1900	4550	7750	11400	15000
Cocoon Production	44374	22700	5841	7021	12613	18613	22700
Raw Silk	4671	2600	584	740	1363	2070	2600
Andhra Pradesh							
Mulberry Plantation	43289	10000	2000	4000	6000	8000	10000
Cocoon Production	19000	11000	2300	3925	6425	6425	11000
Raw Silk	1727	960	233	394	675	675	960
Tamil Nadu							
Mulberry Plantation	31752	11460	1000	2740	5660	8580	11460
Cocoon Production	6721	7937	594	1895	3918	5941	7937
Raw Silk	560	859	25	255	457	659	859
West Bengal							
Mulberry Plantation	14536	7200	1000	2300	3700	5300	7200
Cocoon Production	11323	7627	212	1567	2885	4381	7627
Raw Silk	871	636	18	133	244	374	636
Jammu & Kashmir							
Mulberry Plantation	1541	2750	250	650	1200	1900	2750
Cocoon Production	800	420	-	-	83	221	420
Raw Silk	90	42	-	-	7	21	40
Central Silk Board							
Mulberry Plantation	6630	11200	720	2640	5520	8800	11200

Cocoon Production	1063	8400	540	1980	4140	6600	8400
Raw Silk	106	840	54	200	414	660	840
Grand Total							
Mulberry Plantation	238204	57610	6870	16880	29830	43980	57610
Cocoon Production	83281	58084	9487	16388	30064	44205	58084
Raw Silk	8025	5937	914	1722	3160	4659	5935

Source: Silk man's companion 1992, CSB, Bangalore.

NGOs and women groups is intended basically to promote greater participation of women in sericulture, provide employment to the land-less and other under-privileged groups to the extent possible through sericulture.

The Table 4.12 reveals that under the NSP so far financial assistance has been provided to 122 organisations, of which 103 are in the traditional states and 19 are in the pilot states. The Central Silk Board has undertaken an internal review to assess the qualitative and quantitative impact of the sericulture schemes implemented by the NGOs in the pilot states under NSP. The Central Silk Board has been providing various forms of support to the NGOs to take up sericulture development programmes.

Table 4.12

ASSISTANCE TO NON-GOVERNMENTAL ORGANISATIONS UNDER NSP

(Rs. in lakhs)

State	No. of N.G.Os.	Assistance approved
Karnataka	21	59.09
Tamil Nadu	41	70.80
Andhra Pradesh	31	40.84
West Bengal	10	12.45
Pilot States	19	95.78
Total	122	278.96

Source: Indian Silk, CSB Vol. 33, No. 6, October 1994.

RESEARCH AND TRAINING FOR THE DEVELOPMENT OF SERICULTURE

The efforts made in the fifties to transplant temperate technologies into the predominantly tropical regions of our country failed. Then it was realised that a tropical oriented research relevant to Indian conditions was necessary. To meet this requirement the Central Sericultural Research and Training Institute was established in Mysore in 1962, and the Central Sericultural Research Station at Berhampur was re-organised and up graded as a full-fledged research institute. In 1964, came up the Central Tasar Research Station and soon this was followed by the establishment of the Regional Tasar Research Station on Temperate Oak Tasar in 1972. In the same year the Regional Research Stations in Mirza and Titabai were started for research on muga and eri sericulture respectively, when the State-run stations were taken over by the Central Silk Board.

Then the International Centre for Training and Research in Tropical Sericulture (ICTRETS) was established in Mysore, marking a significant step forward in international co-operation in sericulture development. The ICTRETS was started with the purpose of providing training to candidates from the various developing countries in the tropics in Tropical Sericulture Technology. The Central Silk Technological Research Institute, perhaps the first institute of its kind in the entire world was set up in Bangalore in 1983 to conduct research on silk reeling, spinning, weaving and processing .

These research and training institutes of the Central Silk Board are putting an all out effort on their part to give research support essential to the development of sericulture in the country. The various Regional Research Stations and Research sub-stations attached to the main institutions attend to the problems in thier respective regions and offer tailor-made solutions for them. There is a wide net-work of extension centres in the sericulture states, both mulberry and non-mulberry,which disseminate the laboratory findings and transfer of technology to the farmers so that the progress of sericulture is continued without letup or hindrance.[13]

Training at the junior level ranging from six months to one year is imparted by the sericultural States themselves while the Central Silk Board is responsible for advance training. The CSB offers both National level and International level training courses. The Ministry of Agriculture and Rural Development Department, Government of India has sanctioned a total outlay of Rs 79.54 lakhs for establishing 13 sericulture training schools for farmers.

SERICULTURE DEVELOPMENT IN STATES

In the early days of sericulture in India it was confined to a few states in which the agroclimatic conditions were suitable for its development. With the advent of the Five Year Plans and continuous research work, sericulture which was practiced only in the traditional states of Karnataka, West Bengal, Tamilnadu and Jammu & Kashmir in the early years has now spread to Andhra Pradesh and other non-traditional states. The map 4.1 gives a picture on the spread of sericulture in different states. The combination of factors including a new awareness of the economies of sericulture operations and the thrust given to it by the government agencies backed by multi-lateral aids has resulted in a near uniform spread of the industry in the new areas of the country. The Table 4.13 gives a picture of the raw silk producing states in India.

Table - 4.13

STATE WISE MULBERRY ACERAGE, PRODUCTION OF DFLs, REELING COCOONS AND RAW SILK PRODUCTION (1990-91)

Sl. No.	State	Mulberry Area (in thousand ha.)			DFLs (lakh Nos.) (in hundred tonnes)	Production of reeling cocoons	Raw Silk (in hundred tonnes)
		Rainfed	Irrigated	Total			
1.	Andhra Pradesh	-	76.35	76.35	249.20	322.62	31.94
2.	Assam	1.35	-	1.35	6.63	2.14	0.18
3.	Arunachal Pradesh	0.05	-	0.05	0.02	0.12	0.01
4.	Bihar	4.00	0.08	4.08	23.88	6.48	0.65
5.	Gujarat	-	0.15	0.15	-	0.04	Neg.
6.	Himachal Pradesh	-	0.59	0.59	3.60	0.98	0.08
7.	Jammu & Kashmir	1.80	-	1.80	35.81	7.00	0.18
8.	Karnataka	60.97	88.82	149.79	2112.11	590.33	62.14
9.	Kerala	-	0.62	0.62	-	0.48	0.01
10.	Madhya Pradesh	1.22	-	1.22	10.79	1.40	0.09
11.	Maharashtra	-	2.10	2.10	11.27	2.91	0.06
12.	Manipur	15.20	-	15.20	8.80	2.94	0.23
13.	Mizoram	0.93	-	0.93	0.26	0.08	Neg.
14.	Meghalaya	1.25	-	1.25	3.16	0.09	0.01
15.	Nagaland	0.05	-	0.05	0.41	0.16	0.01
16.	Orissa	1.38	0.03	1.41	1.66	0.30	0.03
17.	Punjab	0.01	0.05	0.06	0.62	0.20	Neg.
18.	Rajasthan	-	0.47	0.47	-	0.13	0.01
19.	Sikkim	0.04	-	0.04	-	Neg.	-
20.	Tamil Nadu	4.86	31.94	36.80	195.00	118.00	10.72

21.	Tripura	0.92	-	0.92		14.22	0.40	107.80	0.02	
22.	Uttar Pradesh	0.23	0.49	0.72		490.00		0.21		
23.	West Bengal	14.98	2.19	17.17	2.86	2.12		8.29		
	Total	109.24	203.88	313.12		3170.30		1166.72		114.87

Source: Silk in India, Statistical Biennial, 1992, CSB, Bangalore.

KARNATAKA

Karnataka popularly called 'Silk Bowl' of India has continued to maintain its pre-eminent position in raw silk production in the country. The entire credit for introducing sericulture into that region goes to Tippu Sultan, a ruler of the then Mysore State in the eighteenth century.

Tippu Sultan's enthusiasm and efforts paved the way for the successful introduction of Sericulture activity around Mysore in the 1780s. Extensive efforts to obtain good silk worm strains and expert sericulturists were made and the Sultan sent artisans to Bengal to learn the techniques of rearing and reeling and took meticulous care in popularising sericulture.

Though Tippu was successful in introducing sericulture, he derived little commercial advantage from it. But support to sericulture was continued even after Tippu Sultan's regime, by the later Mysore rulers and the industry flourished and struck deep roots in Karnataka. By 1941 both cocoons and raw silk could be bought in Ramanagaram. Today Ramanagaram is an important reeling centre and the principal trading centre for silk coons in India.

Presently sericulture is being practiced in about 16,593 villages covering almost all the districts of the state. The Central Silk Board and the State Government have been showing a great attention for developing the sericulture industry in the state. Measures like providing extension support and supplying disease-free layings at subsidised cost etc., have been taken up. Insecurity in sericulture activity due to the effects of diseases and climate has been eliminated and the industry has acquired a sound technological base.[14]

To look after the silk industrial activities and

also to stabilise raw silk prices, a full-fledged Directorate of Sericulture has been established in the state along with the Karnataka Silk Industries Corporation and Karnataka Silk Marketing Board. The National Sericulture Project is under implementation by the Department of Sericulture, Karnataka State since 1989.

Karnataka ranks first in the country in the production of mulberry raw silk. Mulberry cultivation in the state is undertaken both under rainfed and irrigated conditions. The area under mulberry reached 149.79 thousand hectares (47.84 percent of India's total) and raw silk production also rose to 6214 tonnes (54.10 percent of India's total) during 1990-91.

ANDHRA PRADESH

Andhra Pradesh, once a non-traditional state in sericulture, has achieved the distinction of being the second largest mulberry silk producer in the country. The mulberry and raw silk production in the state reached 90.80 thousand hectares (26.61 percent of India's total) and 2,860 tonnes (21.34 percent of India's total), respectively during 1993-94 Mulberry sericulture has become an established industry in the areas adjoining the Karnataka plateau, especially in the districts of Anantapur and Chittoor. It is being practiced fully under irrigated conditions in the state.

A Indo-Swiss mulberry sericulture development project is being implemented in Andhra Pradesh at a total cost of Rs.141.11 lakhs, which includes the Swiss grant to the extent of Rs 131.03 lakhs. The National Sericulture Project is being implemented in the state by the Department of Sericulture.

TAMIL NADU

Tamil Nadu ranks third in the production of mulberry raw silk in India. The industry has struck deep roots in almost all the districts of the state, more particularly in Salem, Dharmapuri, Coimbatore and Maduri. The state has a very good potential in the hilly districts of Nilagiris, Salem and North Arcot for developing bivoltine silkworm rearing. The state produced 1072 tonnes of mulberry raw silk during 1990-91.

The area under mulberry plantation was 36,824 hectares during 1990-91. The expansion in the area under mulberry was possible mainly due to the State Sericulture Department's efforts in implementing the special development programmes.

A mulberry sericulture project with financial assistance from the Swiss Development Co-operation and the National Sericulture Project is being implemented in Tamil Nadu by the Department of Sericulture. Besides the Central Silk Board has been providing infrastructural facilities for the development of sericulture in the state.

WEST BENGAL

West Bengal is the fourth largest mulberry silk producer in the country. Sericulture is a traditional industry in this state and the silk fabrics produced in the Mushirabad District are world famous for their hand woven design. West Bengal produces both mulberry and non-mulberry silks.

Mulberry sericulture is practiced in the Malda, Mushirabad, Birbhum and West Dinajpur districts. The area under mulberry cultivation at the end of 1990-91 was 17,167 hectares and raw silk production was 829 tonnes.

The National Sericulture Project is being implemented in the State by the Department of Sericulture at a total cost of Rs.59.80 crores which includes Rs.32.10 crores of government investment and Rs.17.70 crores credit to beneficiaries through institutional finance for on-farm and silk processing activities. The project envisages expansion of mulberry area by 7000 hectares for increasing raw silk production by 770 tonnes.

JAMMU & KASHMIR

Sericulture is a traditional industry in the State of Jammu & Kashmir. The state is gifted with a climate which is highly congenial for bivoltine silkworm rearing. The area under mulberry plantation during the year 1990-91 was 1941 hectares and the production of mulberry raw silk was 18 tonnes. The sericulture industry is providing at present part-time employment to about 30,000 families in the state, besides full-time employment to 5000 families in different sectors of sericulture activity.

The entire raw silk in the state is produced in government filatures. Both the Central Silk Board and state government have been paying great attention to the development of bivoltine sericulture in the state.

The National Sericulture Project is being implemented in the state by the Department of Sericulture at a total cost of Rs. 25.00 crores which include Rs. 17.70 crores government investment and Rs. 7.50 crores credit

Map 4.1
Sericultural Map of India

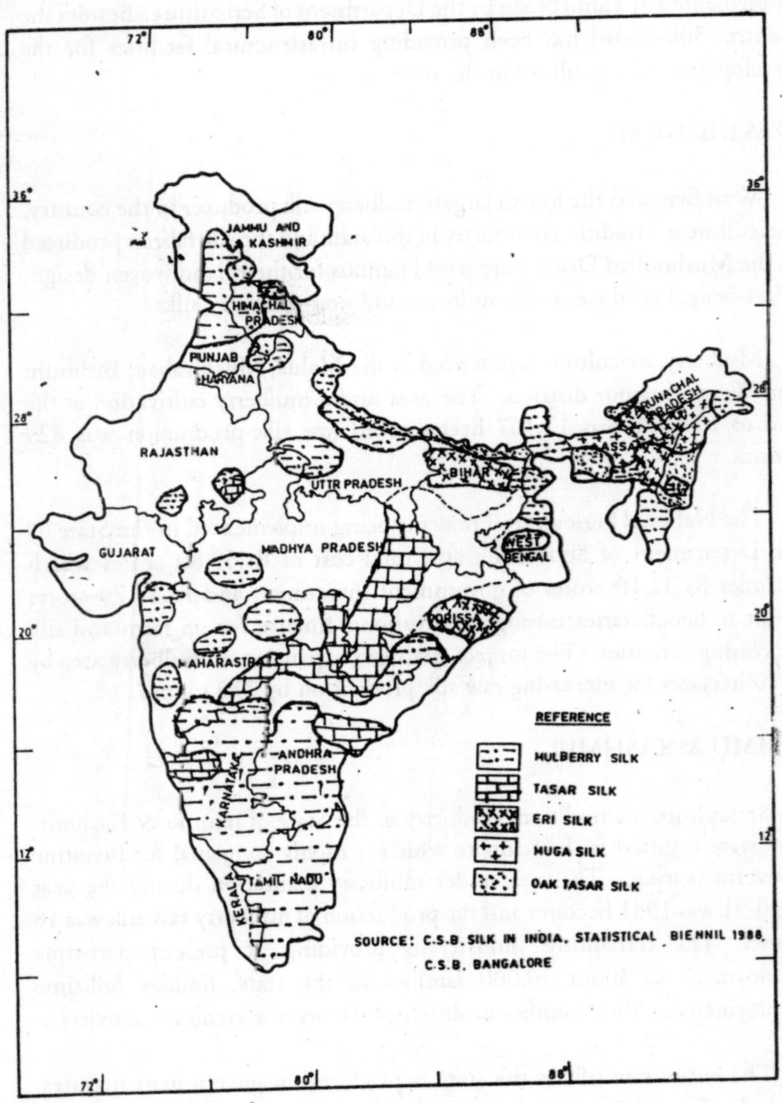

to beneficiaries through institutional finance for on-farm and silk processing activities. The project envisages expansion of mulberry area by 2750 hectares for increasing mulberry raw silk production by 42 tonnes. The Central Silk Board has created additional infrastructure facilities in the state for the development of sericulture.

ASSAM

The famous golden yellow Muga silk is unique and exclusive to India. Muga silk occupies a prominent position in the history and cultural heritage of Assam. Further, Assam accounts for 60 per cent of the total Eri silk production in the country. Mulberry sericulture is practiced in six districts of Assam and there is wide scope for its development. The Central Silk Board is implementing the National Sericulture project, which envisages to cover an additional area of 4000 hectares under mulberry plantation for producing 120 tonnes of mulberry raw silk.

BIHAR

Bihar is one of the leading Tasar silk producing states in the country with 60 per cent of the total tasar production. During 1990-91 the state produced 370 tonnes of raw silk which included 65 tonnes of mulberry raw silk, 281 tonnes of Tasar and 24 tonnes of Eri silk. Under the National Sericulture Project it is estimated to cover an additional area of 2000 hectares under mulberry cultivation for producing 60 tonnes of raw silk.

HIMACHAL PRADESH

Himachal Pradesh is considered as a univoltine state with scattered mulberry plantation of varying stages. Around 600 hectares of mulberry plantation is under the government sector. The Central Silk Board has established research extension centres one each at Bhir and Sirmur for mulberry. The NSP is under implementation in the Solan district of the state.

KERALA

Sericulture was introduced in Kerala very recently. The agro-climatic condition of the state is suitable for the development of sericulture, both in the rain shadow region and the high rainfall areas, where mulberry grows well and silkworms are reared successfully. The State has a congenial climate for development of bivoltine silkworm rearing, specially in the

Western ghats region. The Central Silk Board is implementing the National Sericulture Project in the districts of Palghat and Idukki.

MAHARASTRA

Maharastra is also one among the new states which have taken up sericulture recently. Higher altitude, moderate temperature and good soil conditions in the Western Ghats and Vidharba region of the state, offer good scope for the development of sericulture, especially bivoltine silkworm rearing, in these regions. The area under mulberry cultivation in the state has vast potential for the development of both tasar and mulberry sericulture. The National Sericulture Project is under implementation in the districts of Akola and Bhuldana. Under this project it is envisaged to cover an additional area of 4600 hectares to produce 120 tonnes of raw silk.

ORISSA

Orissa is predominantly a tasar producing state. The agro-climatic conditions prevailing in the state are well suited for tasar, mulberry and eri culture. Mulberry sericulture was introduced recently in the districts of Koraput, Ganjam and Phulbani. The area under mulberry cultivation during 1990-91 was around 400 hectares. The Central Silk Board is implementing the National Sericulture Project in the district of Koraput. Under this project, an additional area of 1600 hectares will be brought under mulberry for an additional production of 120 tonnes of mulberry raw silk.

PUNJAB

Mulberry cultivation is practiced in the sub-mountainous areas of Gurudaspur, Hoshiarpur and Popar districts. The agro-climatic conditions in the state are favourable for mulberry sericulture and during 1990-91 about 1000 families took up sericulture. The Central Silk Board is implementing the National Sericulture Project in the District of Hishiarpur and the project envisages to cover an area of 200 hectares for the production of 30 tonnes of raw silk.

RAJASTHAN

Rajasthan is also a non-traditional state and the climatic conditions in the state are suitable for all the four types of natural silk. During VI Plan period the Central Silk Board and the State Government initiated various measures for developing sericulture in this state. Under the National Sericulture Project an additional area of 400 hectares of mulberry cultivation will be covered for obtaining an additional 60 tonnes of mulberry raw silk.

UTTAR PRADESH

Uttar Pradesh has initiated sericulture schemes recently. Mulberry sericulture is practised both on nature grown mulberry trees and plantations raised on plains and hilly regions. The Central Silk Board is implementing the National Sericulture Project in the districts of Dehradun and Sahranpur. Under this project an additional area of 3,000 hectares will be covered for an additional production of 120 tonnes of mulberry raw silk.

As the focus of the present study is mainly on the states of Andhra Pradesh and Karnataka, a detailed account of the status of sericulture in these states is given in the next chapter.

CONCLUSION

Sericulture in India has an age old history. Some historians hold that mulberry sericulture was first introduced in India, in the 14th and 15th centuries. Since then Indian sericulture has passed through several ups and downs in its journey towards prosperity. Sericulture industry in India owes a great deal to Tippu Sultan, a ruler of the former Mysore State in the 18th century.

After Independence, the Government of India, having recognised the inherent advantages of sericulture, has been giving a prominent place to it in the developmental plans. Through the Five Year Plans the sericulture industry has achieved a tremendous growth. The allocation of funds has gone up from Rs. 60.22 lakhs in the First Five Year Plan to Rs. 31,078.00 lakhs in the Seventh Five Year Plan period. The final outlay proposed for sericulture development during the Eighth Five Year Plan is Rs. 860.76 crores.

In India, mulberry cultivation has been acknowledged as an important crop particularly because of its potential for developing rural areas. As a result the area under mulberry crop which was only 1,70,000 hectares in 1980-81 increased to 3,41,200 hectares in 1993-94. It is employing 60.30 lakh persons in about 59,528 sericulture villages all over the country. The foreign exchange earnings through it were about Rs. 440.53 crores in 1990-91.

The promotional agencies have been playing a key role in the development of sericulture in India. The Central Silk Board, Bangalore and the CSRTI Mysore are helping it by providing suitable technologies, evolving new varieties of mulberry and silk worm races, taking research findings to the farms and giving training to the sericulturists. The biggest impetus in recent times to the development of Indian sericulture is the implementation of the National Sericulture Project (NSP) with Rs. 555 crores financial assistance from the World Bank and the Swiss Development Co-operation. The project is under implementation in five traditional states and 12 non-traditional states of the country.

During early years sericulture was restricted to only a few states. But with the continuous and determined efforts of governmental agencies, now it has spread to new areas all over the country. Presently, it has been practised in 23 states of India, of which only five are traditional states and the remaining 18 are non-traditional states in terms of sericulture and its development.

Note:-

1. Sanjay Sinha, **The Development of Indian Silk,** Oxford and IBH Publishing Co. Pvt. Ltd., New Delhi, P.4.
2. S.R. Charley, **Culture and Sericulture,** Academic Press, London, 1982, P.69.
3. Sanjay Sinha, **The Development of Indian Silk,** Op.cit., P.5.
4. Ibid. P.7.
5. A.R.S. Gopalachar, "March of Sericulture Industry during Post Independence Era", ***Indian Silk,*** Vol.II, No.4, 1972, P.13.
6. CSB, **Silkman's Companion 1992,** Op.cit., P.17.
7. Ibid., PP.17-18.
8. Ibid., PP. 19-26.
9. Ibid., P.13.
10. Ibid., P.9.
11. Manjeet .S. Jolly, **Central Sericultural Research and Training Institute - Its organisational set-up and Achievements,** CSRTI, Mysore, 1981, P.3.
12. G.S. Siva Prakash, "National Sericulture Project", ***Indian Silk,*** Vol. 27, No. 12, April 1989, C.S.B. Bangalore, P.5.
13. CSB, **Silkman's Companion 1992,** Op.cit., PP. 27-28.
14. .Sanjay Sinha, **The Development of Indian Silk,** Op.Cit., p.8.

CHAPTER 5
Growth of Sericulture in Andhra Pradesh and Karnataka States [1]

Sericulture has a long history in Karnataka. More than fifty per cent of India's raw silk is produced in Karnataka. Andhra Pradesh ranks next to Karnataka in mulberry raw silk production. A major portion of India's mulberry raw silk comes from these two states only. Thus sericulture has come to play a unique role in the rural economies of Andhra Pradesh and Karnataka. This chapter makes an attempt to analyse the progress of sericulture, its suitability, concentration and present status in these two states.

PROGRESS OF SERICULTURE IN ANDHRA PRADESH

Andhra Pradesh is the fourth biggest state in India both in terms of area(2,74,400 sq km) and population.[1] It is mainly inhabited by Telugu speaking people. It is bounded by Orissa and Madhya Pradesh on the North, Maharashtra and Karnataka on the west, Tamil Nadu on the South and the Bay of Bengal on the East. It has a long coast line of about 960 km. The State is located between North latitudes 12° - 20° and East longititudes 77° - 85°. It forms the eastern part of the peninsular and South-eastern part of India. It has 23 districts generally divided into three geographical regions known as Circar or Coastal Andhra, Rayalaseema and Telangana.

The climate is generally hot and humid. The state receives its rainfall both from the South-west and the North-east monsoons. The normal rainfall of the state is 890 mm.[2] It varies widely from region to region. It is erratic and inadequate at places and as a result those areas suffer from chronic drought. The Finance and Planning Department of Government of Andhra Pradesh has identified that 100 taluks in 12 districts of the State constitute the hard core of drought-prone areas. It is also identified that 96 per cent of the total population in the Rayalaseema Region is drought affected.[3]

The state is drained by two major river systems i.e., the Krishna and the Godavari. Red soils covering two-thirds of the state are predominant in the Rayalaseema and Telangana districts. The upland portions of the districts of West Godavari, East Godavari, Visakhapatnam and Srikakulam are also covered by red soils. Black soils cover one-fourth of the area in the State.

Agriculture is the main occupation of about 70 per cent of the people in Andhra Pradesh. The state is surplus in food grains, particularly rice. It is popularly known as the Rice Bowl of South India. The other important crops are Jowar, Bajra, Maize, Ragi, Small Millets, Pulses, Tobacco, Cotton, Sugarcane, Groundnut and Bananas. The total food grain production is 123.30 lakh tonnes of which rice constitutes 96.54 lakh tonnes (78.3 per cent). The production of oil seeds is 60.13 lakh tonnes.[4] The other important commercial crop which gained momentum during the last decade especially in the Rayalaseema Region is mulberry.

Andhra Pradesh has now achieved the second place in the production of mulberry raw silk, although it had no place in the sericulture map of India when the State was formed in 1956. The hectarage under mulberry cultivation in 1953 was just two hectares which reached 16 hectares by 1956.[5] Andhra Pradesh was not a traditional state of sericulture and was a new entrant into the sericulture enterprise. But within four decades due to the enormous progress of sericulture in it, the state has achieved the distinction of second place in India's mulberry raw silk production and now it is being treated as one of the traditional states of sericulture in the country.

Palamaner in the Chittoor district was the only sericulture farm formarly. Gradually owing to the efforts of Sir S.V.Ramamurthy adviser to the Governor of the then Madras province, sericulture spread to Hindupur, Araku and other places. Sericulture came to be practiced in a small way with the setting up of a few farms by the Government at Lepakshi, Bhadrachalam, Chintalapudi, Chintalapalli and Palamaneru. However, major development took place only after 1970-71.

Sericulture was originally under the control of the Director of Industries and the apex administrator was a joint Director who dealt with sericulture and other Cottage Industries. Sericulture was transferred to the Director of Handlooms and Textiles in 1967. Keeping in view its importance and potential for development, a separate department of Sericulture was formed by the state government with effect from 20-4-1981.[6]

The entire State was divided into convenient regions for the purpose of smooth administration by the Department of Sericulture. Each region consists of a district or a group of districts. A Deputy Director or an Assistant Director is placed in charge of each region depending on the scale of sericulture. The State Government has set up a Federation of Sericulture and Silk Weavers Co-operative Societies with Hyderabad as its head quarters. The Director or a person nominated by the Government is its chairman and a technical expert is appointed as Managing Directcr. The government nominates the Board of Directors to run the Federation. The Federation has been authorised in a very broad way to carry out any work that would promote sericulture development.

MULBERRY CULTIVATION

A phenomenal increase took place in the hectarage under mulberry cultivation from 2,610 to 90,800 hectares between 1970 - 71 and 1993 - 94. It is interesting to note that all the area of mulberry is under irrigated conditions. However, the Central Silk Board has recently evolved a rainfed variety of mulberry viz. S_{13} which is under trial on an experimental basis in some parts of the state like Paderu, Chintalapalle, Araku Valley etc in Costal Andhra where the rainfall is considerable. The particulars of area under mulberry cultivation in Andhra Pradesh from 1980-81 to 1993-94 is furnished in Table 5.1

Table 5.1

AREA UNDER MULBERRY CULTIVATION IN ANDHRA PRADESH (1980-81 - 1993-94)

Year	Area Under Mulberry (Thousand Hectares)	Percentage to India's total
1980-81	18.90	11.12
1981-82	25.82	14.35
1982-83	33.56	17.05
1983-84	36.95	17.86
1984-85	40.46	18.53
1985-86	45.37	20.83
1986-87	44.15	19.23
1987-88	49.59	20.53
1988-89	56.03	20.90
1989-90	68.29	23.21
1990-91	74.31	23.73
1991-92	80.60	24.33
1992-93	86.02	25.14
1993-94	90.80	26.61

Source : Central Silk Board, Bangalore.

It is clear from the Table 5.1 that the area under mulberry cultivation in the state has registered a remarkable growth. Every year there was an increase in the area under mulberry cultivation except in 1986-87 when there was a relative decrease in it. Severe drought conditions and depletion of ground water levels were the probable causes for this relative decrease during that year. By the end of 1993-94 an extent of 90.80 thousand hectares of land was brought under mulberry cultivation. The increase was mainly due to conversion of the land under food crops and a few commercial crops for mulberry cultivation because of its various advantages.

The table (No. 5.1) also shows the contribution of Andhra Pradesh to India's total mulberry hectarage. The share of Andhra Pradesh was insignificant during 1980-81 (11.12 per cent) and by the end of 1993-94 its contribution increased to 26.61 per cent. The growth of area under mulberry and its percentage to India's total is also shown in Graph No. 5.1. The growth rate of area under mulberry in Andhra Pradesh between 1980-81 and 1993-94 is presented in Table 5.2.

Table 5.2

GROWTH RATE OF AREA UNDER MULBERRY CULTIVATION IN A.P (1980-81 - 1993-94)

Category	Constant	Regression Co-efficient	Growth Rate	R^2	t-Value
Area Under Mulberry	3.0661	0.110	11.609	0.958	16.575

Source: Computed from the data drawn from Table 5.1.

In Table 5.2 the coefficient of area under mulberry cultivation is significant at 1 per cent level indicating that the coefficient of determination R^2 (0.958) and the growth rate are significant. The above results reveal that the area under mulberry cultivation in Andhra Pradesh increased at the rate of 11.609 percent during the period 1980-81 to 1993-94.

The concentration of the area under mulberry is mainly in the four districts of Rayalaseema, because of favourable climatic conditions prevailing in this region for mulberry cultivation. Regionwise area under mulberry over a decade from 1984-85 to 1993-94 is given in Table 5.3.

Table 5.3 shows that the land under mulberry cultivation is concentrated mainly in the Rayalaseema region of the state. The other two regions Coastal and Telangana, project a nominal picture. It may also be observed that the share of Rayalaseema which was 90.57 per cent in 1984-

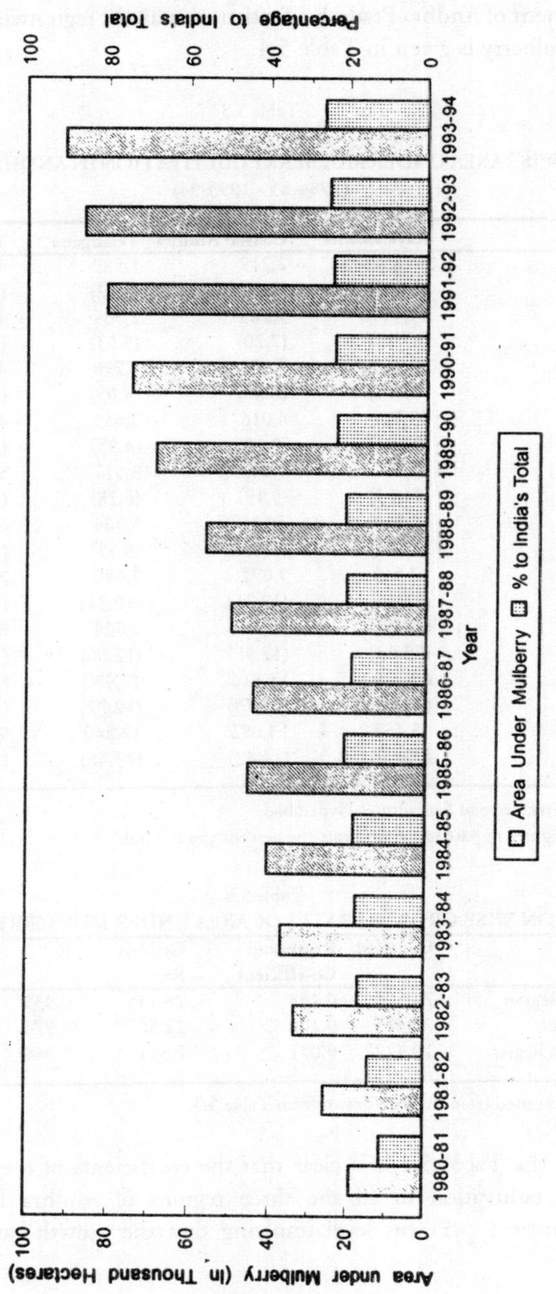

Graph 5.1
Area Under Mulberry Cultivation in A.P. (1980-81 - 1993-94)

85 came down gradually to 72.10 per cent in 1993-94. It implies that the share of the other two regions has increased which may be due to the realisation of the benefits of this crop by the farmers of those regions. However, Rayalaseema still forms the major force in the sericulture development of Andhra Pradesh. Statistical analysis, regionwise of the area under mulberry is given in Table 5.4.

Table 5.3

REGION WISE AREA UNDER MULBERRY CULTIVATION IN ANDHRA PRADESH (1984-85 - 1993-94)

Year	Rayalaseema	Coastal Andhra	Telangana	Total
1984-85	36,652	2,219	1,589	40,460
	(90.57)	(5.48)	(3.92)	(100.00)
1985-86	39,359	3,265	2,554	45,366
	(87.17)	(7.20)	(5.63)	(100.00)
1986-87	39,359	3,003	1,790	44,152
	(89.15)	(6.80)	(4.05)	(100.00)
1987-88	43,121	4,016	2,453	49,590
	(86.91)	(8.10)	(4.95)	(100.00)
1988-89	47,370	5,147	3,517	56,034
	(84.53)	(9.19)	(6.28)	(100.00)
1989-90	55,613	7,111	5,564	68,288
	(81.44)	(10.41)	(8.15)	(100.00)
1990-91	57,549	9,072	7,688	74,309
	(77.40)	(12.21)	(10.34)	(100.00)
1991-92	60,476	10,460	9,664	80,600
	(75.03)	(12.98)	(12.06)	(100.00)
1992-93	63,220	11,816	10,990	86,020
	(73.50)	(13.73)	(12.77)	(100.00)
1993-94	65,452	13,082	12,240	90,774
	(72.10)	(14.42)	(13.46)	(100.00)

Source : Directorate of Sericulture, Hyderabad.

Note : Figures in parentheses indicate the percentages to total.

Table 5.4

REGION WISE GROWTH RATES OF AREA UNDER MULBERRY IN A.P.

Category	Constant	Regression Co-efficient	Growth Rate	R^2	t-Value
Telangana Region	7.0506	0.248	28.135	0.939	11.361
Costal Region	7.5415	0.206	22.867	0.974	17.261
Rayalaseema Region	10.4227	0.071	7.381	0.966	15.010

Source: Computed from the data drawn from Table 5.3.

From the Table 5.4 it is clear that the coefficients of the area under mulberry cultivation in all the three regions of Andhra Pradesh are significant at 1 per cent level implying that the growth rates and the correlation coefficient are significant. Of the three regions Telangana has the highest growth rate i.e. 28.135 per cent, and the lowest, 7.38 per cent, is in Rayalaseema, during the period form 1984-85 to 1993-94.

COCOON PRODUCTION

With the increase in area under mulberry cultivation the production of reeling cocoons has also increased substantially. The particulars of production of reeling cocoons in Andhra Pradesh since 1980-81 are given in Table 5.5.

Table 5.5 depicts that the production of reeling cocoons in Andhra Pradesh faced frequent fluctuations during the period from 1980-81 to 1985-86. Only after 1985-86 there was a steady progress in it and it reached 24.51 thousand tonnes by the end of 1993-94. It is observed that poor quality of layings supplied by the private grainages, lack of timely supply of good quality layings from the government grainages, use of old and outdated rearing appliances and adverse climatic conditions were mainly responsible for the fluctuations in the production of reeling cocoons. The status of cocoon production during the above period is also shown in Graph 5.2.

Table 5.5

COCOON PRODUCTION IN ANDHRA PRADESH (1980-81 - 1993-94)

Year	Cocoon Production in A.P.	
	Quantity (in Thousand Tonnes)	Percentage to India's total
1980-81	10.98	18.86
1981-82	7.87	14.25
1982-83	12.01	17.98
1983-84	13.05	18.31
1984-85	10.79	14.41
1985-86	12.36	16.11
1986-87	14.92	18.29
1987-88	19.05	22.01
1988-89	24.58	25.48
1989-90	28.17	25.51
1990-91	32.26	27.65
1991-92	27.00	25.19
1992-93	34.52	26.62
1993-94	24.51	19.90

Source : Directorate of Sericulture, Hyderabad.

The trends in cocoon production in Andhra Pradesh over a period of 14 years is statistically analysed and presented in Table 5.6

Graph 5.2

Cocoon Production in Andhra Pradesh, (1980-81 1993-94)

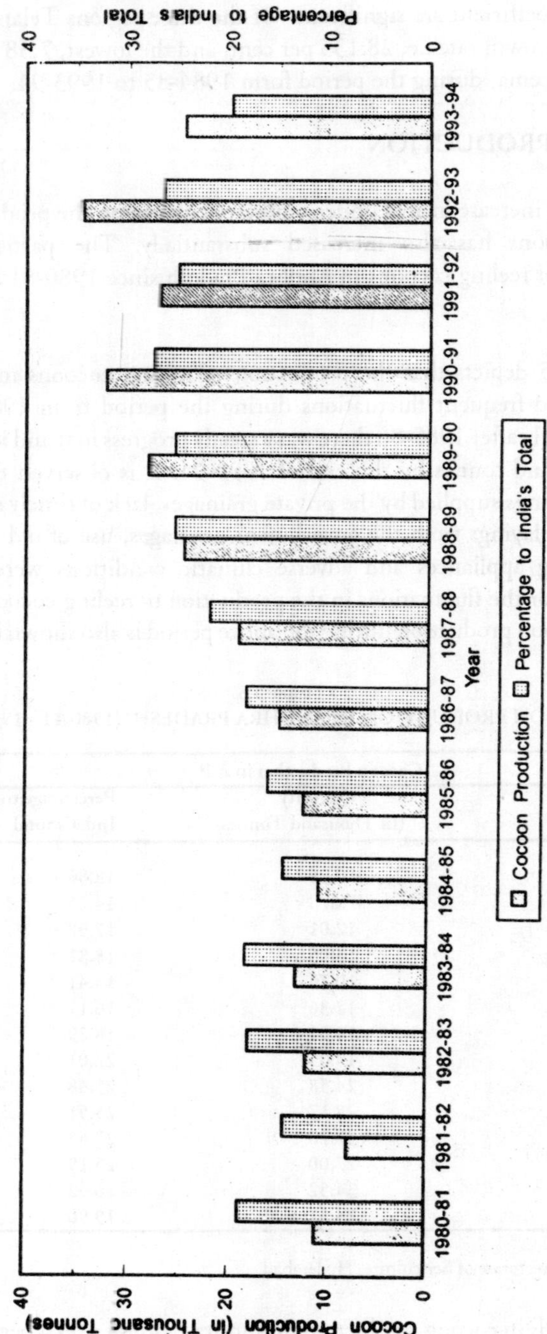

Table 5.6

GROWTH RATE OF COCOON PRODUCTION IN A.P.

Constant	Regression Coefficient	Growth Rate	R^2	t- value
2.1737	0.096	10.028	0.833	7.739

Source: Computed from the data drawn from Table 5.5.

It is evident from Table 5.6 that the growth rate of cocoon production in Andhra Pradesh was 10.028 per cent per year during the period from 1980-81 to 1993-94. The coefficient of determination R^2 (0.833) is significant at 1 per cent level implying that there was a significant growth in the production of cocoons.

RAW SILK PRODUCTION

As a result of the steady progress in the field of cocoon production the quantity of raw silk produced in the state also showed an upward trend. The yearwise production of raw silk and its contribution to India's total is given in Table 5.7.

Table 5.7

TRENDS IN RAW SILK PRODUCTION IN ANDHRA PRADESH (1980-81 - 1993-94)

Year	Raw Silk Production (in Thousand Tonnes)	Percentage to India's Total
1980-81	0.08	17.43
1981-82	0.56	11.67
1982-83	0.75	14.39
1983-84	0.90	15.85
1984-85	1.25	18.12
1985-86	1.03	14.65
1986-87	1.46	18.48
1987-88	1.39	22.34
1988-89	2.59	26.76
1989-90	2.79	25.57
1990-91	3.19	27.76
1991-92	2.85	26.74
1992-93	3.14	24.15
1993-94	2.86	21.34

Source : Central Silk Board, Bangalore.

From Table 5.7 it is evident that the production of raw silk shows a fast growth in Andhra Pradesh. The production was only 0.08 thousand tonnes

Graph 5.3
Raw Silk Production in Andhra Pradesh (1980-81 - 1993-94)

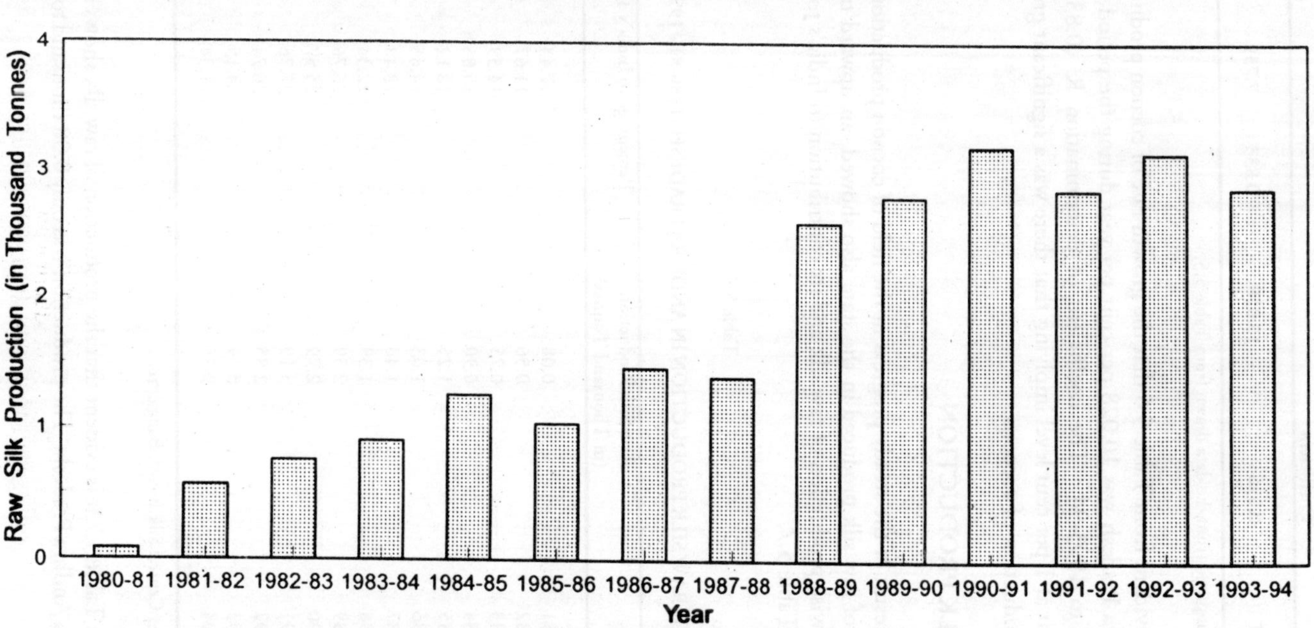

Map 5.1
Spread of Mulberry Sericulture in Andhra Pradesh

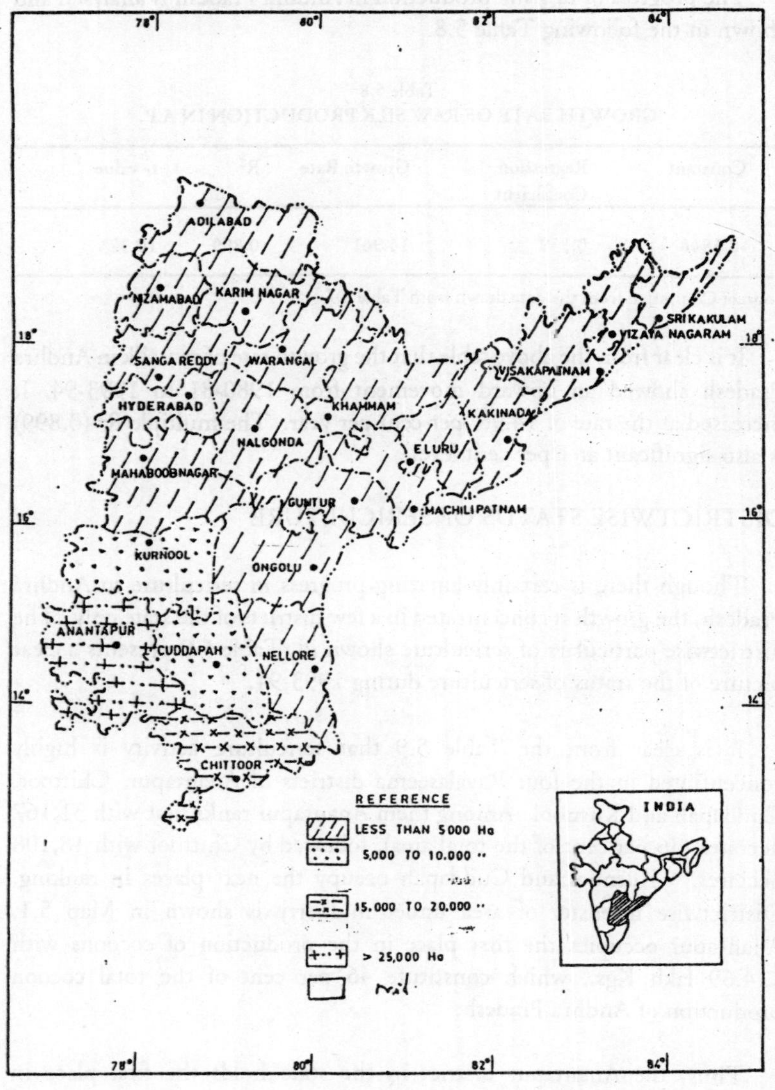

in 1980-81 and by the end of 1993-94 it reached to 2.86 thousand tonnes (Graph 5.3). The share of Andhra Pradesh in India's total production of raw silk also rose from 17.43 to 21.34 per cent during this span of 14 years.

The progress of raw silk production in Andhra Pradesh is analysed and shown in the following Table 5.8.

Table 5.8
GROWTH RATE OF RAW SILK PRODUCTION IN A.P.

Constant	Regression Coefficient	Growth Rate	R^2	t- value
-0.5848	0.139	14.961	0.899	10.325

Source: Computed from the data drawn from Table 5.7

It is clear from the above table that the growth rate of raw silk in Andhra Pradesh showed an upward movement from 1980-81 to 1993-94. It increased at the rate of 14.96 per cent per year. The multiple R^2 (0.899) is also significant at 1 per cent level.

DISTRICTWISE STATUS OF SERICULTURE

Though there is certainly amazing progress in sericulture in Andhra Pradesh, the growth is concentrated in a few districts of the state only. The districtwise particulars of sericulture shown in Table 5.9 presents a clear picture of the status of sericulture during 1993-94.

It is clear from the Table 5.9 that sericulture activity is highly concentrated in the four Rayalaseema districts of Anantapur, Chittoor, Cuddapah and Kurnool. Among them Anantapur ranks first with 31,167 hectares (36 per cent of the total area), followed by Chittoor with 18,108 hectares. Kurnool and Cuddapah occupy the next places in ranking. Districtwise intensity of area under mulberry is shown in Map 5.1. Anantapur occupies the first place in the production of cocoons with 114.69 lakh Kgs., which constitute 46 per cent of the total cocoon production of Andhra Pradesh.

Thus, the Anantapur district in the state holds the first place in sericulture activity. This is due to the favourable climatic conditions in this district and also the rising interest of farmers in this very lucrative activity.

Table 5.9

DISTRICT WISE PARTICULARS OF SERICULTURE IN ANDHRA PRADESH (1993-94)

S. No.	District	Area Under Mulberry Cultivation	Cocoon Production (InLakh Kgs.)
1.	Anantapur	33,190	114.69
2.	Chittoor	18,108	80.79
3.	Cuddapah	6,983	9.65
4.	Kurnool	7,163	10.06
5.	Prakasam	2,986	3.54
6.	Nellore	1,158	3.32
7.	Guntur	1,263	1.03
8.	Krishna	1,646	3.55
9.	E.Godavari	1,433	4.19
10.	W.Godavari	1,476	3.21
11.	Visakhapatnam	1,200	0.35
12.	Vijayanagaram	1,163	0.30
13.	Srikakulam	773	0.69
14.	Mahaboobnagar	1,284	0.61
15.	Ranga Reddy	897	0.19
16.	Nalgonda	1,150	3.87
17.	Karimnagar	2,152	0.40
18.	Nizamabad	513	0.37
19.	Adilabad	794	0.84
20.	Warangal	2,050	1.68
21.	Khammam	1,780	0.25
22.	Medak	1,612	2.43
23.	Hyderabad	--	--
	Total	90,774	245.110

Source: Directorate of Sericulture, Hyderabad.

MARKETING OF REELING COCOONS

The transactions made at the government cocoons markets and the revenue obtained through the sale of cocoons are shown in Table 5.10.

Table 5.10 shows that the quantity of cocoons transacted in the government cocoon markets was low when compared to the total production in the state. Hence, there is the need to evaluate the functions of government cocoon markets, their approach to the sericulturists and the increasing interference of middlemen in the transactions. As the majority of sericulturists are illiterate and transport facilities from the remote villages are not available, the business of selling cocoons had to be completed in the local unorganised cocoon markets. In addition the risk and strain involved in transporting cocoons to far away government cocoon markets forced them to complete their transactions at the unorganised

Table 5.10

TRANSACTIONS MADE IN GOVERNMENT COCOON MARKETS IN ANDHRA PRADESH,
(1986-87 - 1991-92)

Year	Quantity of Cocoons Market Fee	Marketed (in tonnes) (Rs. in lakhs)	Total Value Realised Collected (Rs. in lakhs)
1986-87	4035	1604.91	33.38
1987-88	3769	2016.21	40.27
1988-89	3973	2576.17	51.52
1989-90	5703	4023.20	79.92
1990-91	6247	4351.20	87.18
1991-92	5900	6695.00	133.92

Source : Directorate of Sericulture, Hyderabad.

local markets. After harvesting if the cocoons remain unsold for two or three days worms will develop and come out of the cocoons breaking them open. Then the cocoons become waste and cannot be used.

But some times sericulturists of Andhra Pradesh are seen taking their cocoons to the cocoon markets in the adjacent Karnataka State to get profitable prices for their produce. However, there has been a raise, though slow, in the cocoons transacted in the government cocoon markets in recent years.

The growth rate of transactions made at the government cocoon markets is presented in Table 5.11.

It is clear from the Table 5.11 that the transactions made at the government cocoon markets in Andhra Pradesh has been increasing at 11.395 per cent per year. The multiple correlation coefficient ($R^2 = 0.767$) is significant at 1 per cent level indicating that the growth rate is significant.

Table 5.11

GROWTH RATE OF THE TRANSACTIONS MADE IN GOVERNMENT COCOON MARKETS IN A.P.

Constant	Regression Coefficient	Growth Rate	R^2	t- value
8.1049	0.108	11.395	0.767	3.625

Source: Computed from the data drawn from Table 5.10

INFRASTRUCTURAL FACILITIES

The development of any industry mainly depends on the available infrastructural facilities. In this context it is necessary to review the infrastructural facilities available in Andhra Pradesh for Sericulture development. Infrastructural facilities in Andhra Pradesh were very inadequate because sericulture was not a traditional crop. As it is a new enterprise to the farmers in the state it took some time to build up infrastructure facilities. Since 1974 with the inception of Drought Prone Areas Programme there has been a rise in these facilities in the state. The infrastructural facilities available in the state up to the end of July, 1991 are shown in Table 5.12.

Table 5.12

INFRASTRUCTURAL FACILITIES IN ANDHRA PRADESH (JULY 1991)

S.No.	Infrastructure Available	Number
1	Government Silk Farms	44
2	Government Grainages	44
3	Technical Service Centres	125
4	Government Chawkie rearing Centres	69
5	Private Chawkie rearing Centres	202
6	Cocoon Markets	27
7	Sericulturists-cum-farmers Co-operatives	24
8	Silk Exchange	1
9	Government Twisting Units	10
10	Seed Farms	95
11	Reeling Units	61
12	Semi-automatic reeling units	1
13	Regional Training Centres	3

Source : Annual Reports, Directorate of Sericulture, Hyderabad.

PROGRAMMES AND PROJECTS IN ANDHRA PRADESH

The status of sericulture in a given area is also dependent on the various schemes and projects implemented there by the government. Hence, it is also necessary to have a review of the various schemes and projects responsible for the accelerated development of sericulture.

NATIONAL SERICULTURE PROJECT (NSP)

National Sericulture Project with World Bank assistance, with a total outlay of Rs.118.44 crores, including credit for the on-farm and non-farm sectors was sanctioned for implementation in Andhra Pradesh over a period of Five Years commencing from 1989-90. The principal objective of the project was to increase silk yarn production to the level of 2,500 tonnes per annum within the state. To achieve this objective, an additional area of 10,000 hectares was proposed to be brought under mulberry cultivation during the project period.[7]

The project proposed to generate additional employment to about 1.62 lakh persons. It envisaged to provide full-time employment and income to those belonging to the Scheduled Caste and Scheduled Tribes and also other weaker sections in the state. The break up of the sanctioned amount of 118.44 cores is given in Table 5.13.

For the first three years of the project (1989-90 to 1991-92) a target of 6073 hectares was fixed to be brought under mulberry cultivation. Actually an extent of 26,606 hectares was achieved which is more than four times than the target fixed. In addition there was considerable progress in the non-farm sector also.[8]

Table 5.13
DETAILS OF FINANCIAL OUTLAY UNDER NSP

Description		Amount (Rs. in lakhs)
Infrastructural and Equity		
(World Bank Aid)		32.20
Credit form NABARD		19.34
Subsidies		
1. I.R.D.P. Subsidy	3.90	
2. State Government Subsidy	3.34	
3. C.S.B. Subsidy for Bivoltine Cocoon/Silk	1.89	9.13
Credit from I.D.B.I		22.16
Own funds of entrepreneurs		10.55
Total		93.89
Weaving/Printing and Dyeing units under General line of credit through IDBI 24.55		
Total		118.44

Source : Directorate of Sericulture, Hyderabad

INDO-SWISS PROJECT

A project with Rs.242.47 lakhs with the assistance of Switzerland Government through the Swiss Development Co-operation, was sanctioned in 1987-88 in Andhra Pradesh. The main objective of the project was to improve the status of the small and marginal farmers engaged in sericulture.

The project also envisaged to bridge some of the missing linkages within the sericulture industry and to provide direct benefit to the sericulturists. The programme included the construction of (1) Chawkie rearing centres, (2) mobile disinfection units, (3) infrastructure facilities for silk reeling and (4) providing training to silk reelers. In addition to the above, it also included provision of facilities for marketing cocoons, establishment of Dupion Silk Units and production of bivoltine yarn.[9]

The Swiss Development Co-operation released an amount of Rs.85.00 lakhs for implementation of the scheme, during 1987-88 and 1988-89. The first phase of the project was completed in March 1991.

Thirty hatching centres were established in eight districts. A quantity of 16,71,186 dfls were brushed covering 7,653 sericulturists in farmers hatchery centres. Under the scheme "Mobile Disinfection Units", 21,231 rearing sheds were disinfected, benefitting 21,185 sericulturists. Infrastructure facilities were provided to 151 reelers in the state. Training was imparted to 466 persons in modern techniques of silk reeling for establishment of their own units.[10]

The second phase of the Indo-Swiss Project commenced in 1991-92 for a period of three years. The total project cost was Rs.320.97 lakhs of which these Swiss Development Co-operation's contribution was Rs.205.87 lakhs.

As envisaged in the project, 174 candidates underwent training in silk reeling techniques and 160 Departmental staff were trained in management techniques at the National Institute of Rural Development, Hyderabad under the organisational development component.[11] Nurseries were raised to establish tree plantation. Drip irrigation system was provided.

In addition to the external aided projects mentioned above there were a number of other governmental programmes of which the Drought Prone Areas Programme (DPAP) and the Integrated Rural Development Programme (IRDP) contributed much for the development of sericulture in Andhra Pradesh.

DROUGHT PRONE AREAS PROGRAMME

The Drought Prone Areas Programme was started on the basis of the mid-term appraisal of the Fourth Plan. The Programme was confined to those areas which had been originally taken up under the Rural Works

Programme in 1970-71. The strategy of the programme was to maximise the production in the years when rainfall was good, and minimise the losses when rainfall was not adequate. Under this programme, funds were liberally allotted for sericulture development in Andhra Pradesh until 1988-89. From the year 1990-91 amount was provided for the supply of high-yielding variety of mulberry cuttings for the new farmers to bring an additional area under mulberry cultivation in the districts of Anantapur, Chittoor, Kurnool, Cuddapah, Prakasam, Ranga Reddy and Nalgonda.[12]

INTEGRATED RURAL DEVELOPMENT PROGRAMME (IRDP)

The IRDP was launched in 1978-79. The programme was restructured to meet the goals, like (1) growth and production, (2) benefits to the identified target groups in the dis-advantaged sections of the rural community and (3) full employment within a certain time-frame. The target group consisted of small and marginal farmers, landless labourers and rural artisans.[13]

From 1980-81, Andhra Pradesh started securing help and financial assistance from the IRDP. During the year 1990-91 under the IRDP loaning programme, 8,210 units were covered as against the target of 12,347 units. A sum of Rs.278.407 lakhs was released towards subsidy as against the outlay of Rs.384.78 lakhs. Under the infrastructure scheme of this programme, an amount of Rs.15.53 lakhs was spent as against the total allocation of Rs.23.63 lakhs.

S.C. ACTION PLAN

A special programme with grants from the S.C. Action Plan, has been under implementation for the benefit of the Scheduled Caste beneficiaries. The Pattern of assistance is, subsidy 30 per cent, marginal money 20 per cent, and loan component 50 per cent on the total unit cost Rs.19,800 per 0.40 hectares. During 1990-91 an area of 1505 hectares was brought under mulberry by the Scheduled Caste beneficiaries bringing the total area under mulberry to 7366 hectares from the inception of the programme.[14]

TRIBAL SUB-PLAN

Under this programme, each Scheduled Tribe beneficiary is eligible to avail himself of 50 per cent subsidy (Subject to a maximum of Rs.5,000), 20 per cent margin money, and 30 per cent loan on the total unit cost of Rs.19,800 per 0.40 hectare. Under this programme 472 hectares were brought under mulberry cultivation by the Scheduled Tribe beneficiaries in sub-plan areas during 1990-91.[15]

In addition to the above programmes there are various other programmes which are also responsible for the planned development of sericulture in the state. They include Special Employment Scheme, Social Forestry, Integrated Tribal Development Agency, Drought Relief Works, Water Shed Area Programme, etc. The state government is also encouraging private entrepreneurs to take up reeling and twisting units.

PROGRESS OF SERICULTURE IN KARNATAKA

The Karnataka State, inhabited predominantly by Kannada speaking people, is situated in the west-central part of Peninsualr India. It consists of a narrow elongated belt between the Arabian Sea and the Western Ghats with a strikingly exquisite and enchanting coast-line of about 400 Km.

The state is situated between 11° - 19°. North latitudes and 74° - 79° East longitudes. It is bounded by the Maharashtra state on the north, Goa on the north-west, Andhra Pradesh on the east, Tamil Nadu on the south and south-east, Kerala on the south-west and the Arabian sea on the west. The new united Mysore state was created in 1956 and was renamed as Karnataka in 1973. The Karnataka state covers an area of 1,92,204 Sq.Km. and occupies 5.35 per cent of the total geographical area of India and ranks seventh among the major states of the country. The state has been divided into 20 districts. As per the 1991 census the states population is 4.50 cores and forms 5.3 per cent of India's total population.

Karnataka has a climate, ranging from the very moist rainy monsoon climate in the west coast, the Western Ghats and Malnad areas to the semi-arid climate in the interior central and northern districts and the arid very warm climate in the Bellary-Bijapur region. It receives rainfall both from South-west and North-east monsoons. The normal rainfall of the state is 1139mm. The annual rainfall is highest over the Western Ghats and lowest in the Eastern Ghats. The state is drained by the east-flowing Krishna, Godavari, Cauvery, and the West-flowing rivers, Pennar and Palar river systems.

Black soils occur in parts of the Bidar, Gulbarga, Bijapur and Belgaum districts. Red sandy soils occur in parts of the Dharwad, Bellary, Chitradurga, Shimoga, Bangalore, Tumkur, Hasan, Dakshina Kannada, Mysore and Mandya districts. Mixed soils occur in parts of the Bijapur,

Dharwad, Raichur, Bellary, and Chitradurg districts. Alluvial soils occur in the Kodagu, Shimoga, Dharwar, Uttara Kannada and Dakshina Kannada districts.

Like Andhra Pradesh Karnataka is also a predominantly agricultural state with 69.1 per cent of the total population living in rural areas mostly dependent on agriculture.[16] The average size of land holding is 2.13 hactres. The net area sown is 107.09 lakh hectares of which 21.6 per cent is under irrigation. The total cropped area in the state is 123.93 lakh hectares (62.4 per cent of total geographical area). The important crops grown in the state are Paddy, Jowar, Ragi, Maize, Bajra, Wheat, Sugarcane, Tobacco, Groundnut, Coconut, Oil seeds and Mulberry.

Sericulture is well-known to Karnataka since the eighteenth century. Its beginnings in the state are traced back to times of Tippusultan, the ruler of the erstwhile Mysore State, who organised a silkworm rearing unit in the southern parts of his dominion. He even sent emissaries to different parts of the world to search for silkworm eggs and finally procured them in South China. As the climatic conditions of Karnataka are favourable for the luxuriant growth of mulberry and rearing of silkworms, sericulture has made enormous progress in this state.

Karnataka takes the first place in the country in countributing to the total raw silk production, accounting for 8,250 tonnes in 1993-94. Both in terms of employment and mulberry cultivation, Karnataka's share in the country exceeds 50 per cent of the total figures of the Nation. The state has the distinction of producing the highest quantity of cocoons every year and during 1993-94 it produced 70,210 tonnes of cocoons. The physical growth of the industry in the state from 1949-50 to 1993-94 is given in Table 5.14. It is clear that the industry grew by leaps and bounds after 1969-70. The area under mulberry cultivation increased from 32.39 thousand hectares in 1949-50 to 160.84 thousand hectares in 1993-94. Cocoon production rose to 70.21 thousand tonnes in 1993-94 where as it was 9.09 thousand tonnes in 1949-50. There has been considerable growth in raw silk production also which went up to 8.25 thousand tonnes in 1993-94 from 0.64 thousand tonnes in 1949-50. The reasons for this remarkable progress are (a) effective extension service rendered by the departmental personnel, (b) organised system of cocoon markets (c) establishment of silk exchanges where buyers and sellers could transact by open auction.

Graph 5.4

Area Under Mulberry in Karnataka (1980-81 - 1993-94)

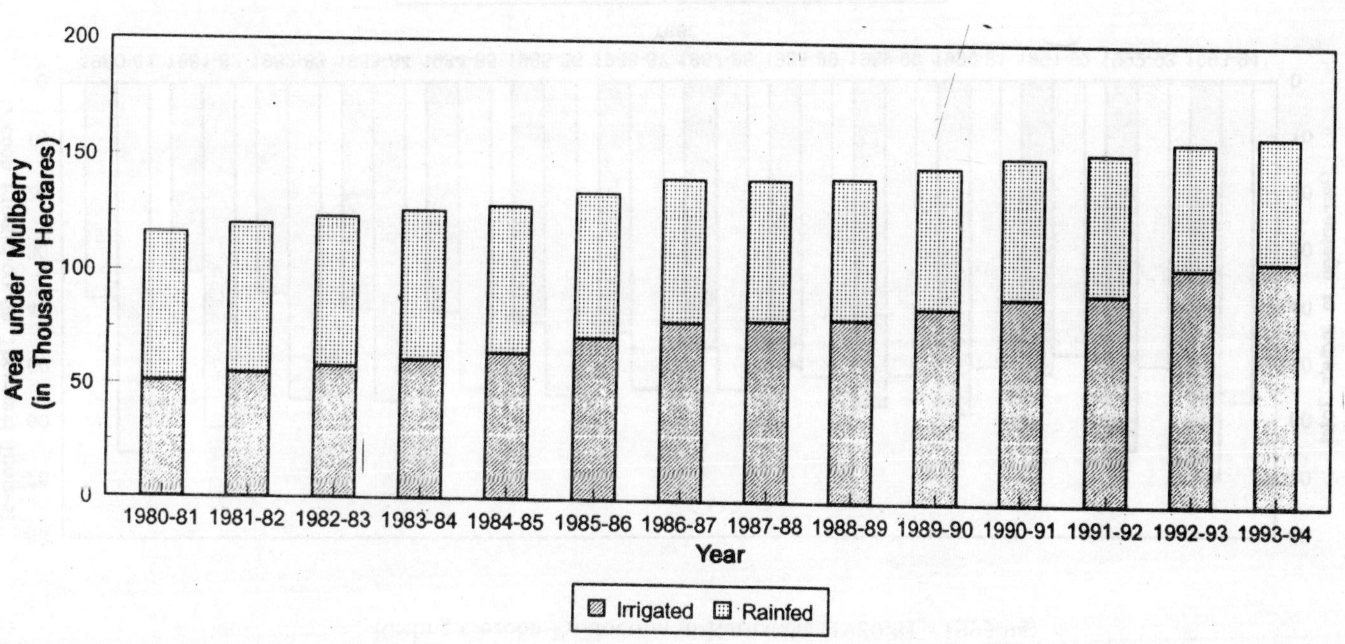

Graph 5.5

Reeling Cocoon Production in Karnataka (1980-81 - 1993-94)

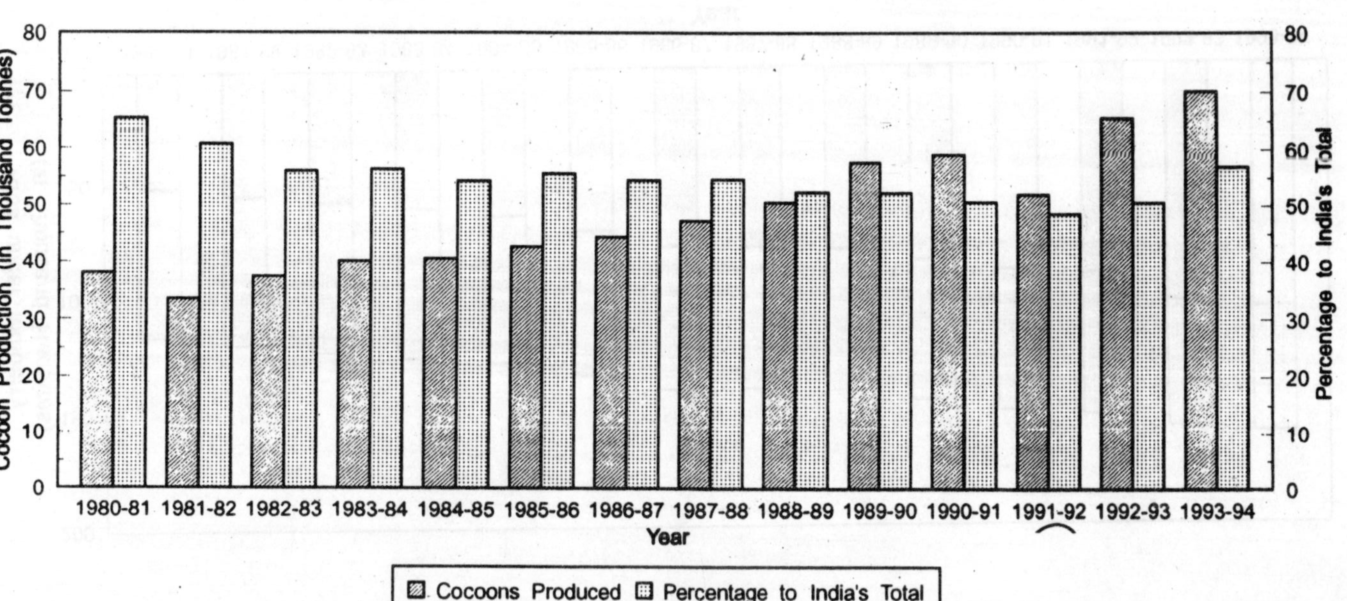

Graph 5.6

Trends in Raw Silk Production in Karnataka (1980-81 - 1993-94)

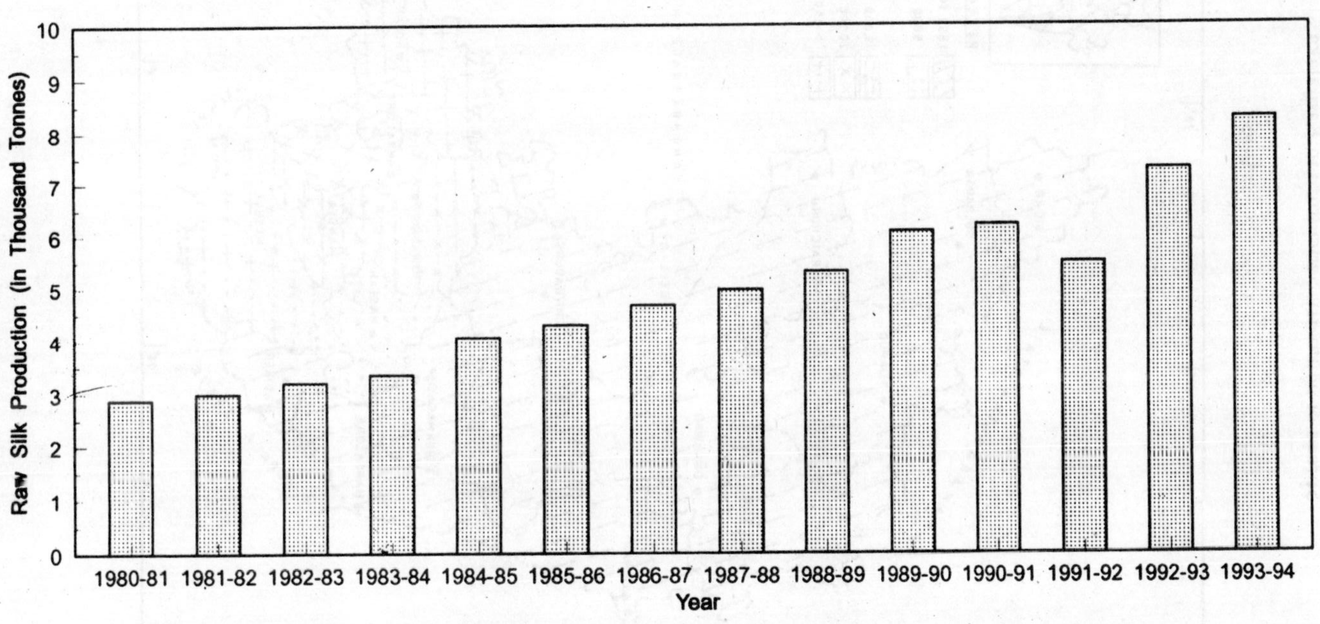

Map 5.2
Spread of Mulberry Sericulture in Karnataka

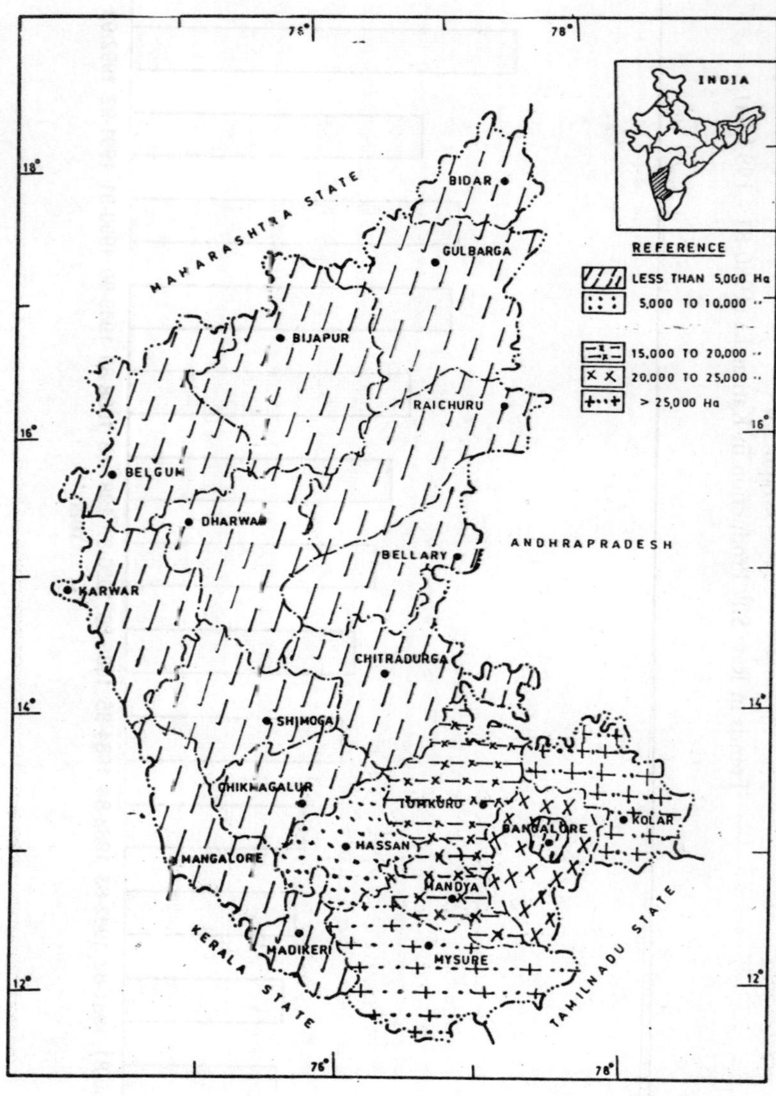

Table 5.14

PHYSICAL GROWTH OF SERICULTURE IN KARNATAKA (1949-50 - 1993-94)

Item	1949-50	1969-70	1993-94
Mulberry Cultivation (in Ha)			
a) Rainfed	8,664	18,524	54,500
b) Irrigated	23,725	67,207	1,06,340
c) Total	32,389	85,830	1,60,840
Cocoon Production (in tonnes)	9,091	21,366	70,210
Raw Silk Production (in tonnes)	636	1,460	8,250
Cocoons per hectare (in Kg.)	281	249	436
Silk per hectare (in Kg.)	19	17	51

Source : Directorate of Sericulture, Bangalore.

MULBERRY CULTIVATION

Mulberry cultivation has become an important commercial crop in the state, and is being cultivated both under rainfed and irrigated conditions. Rainfed mulberry was grown in more areas up to 1983-84 but by the end of 1984-85 irrigated mulberry caught up with it. Since then irrigated mulberry has become more popular and remained ahead of rain-fed mulberry. The area under it has been increasing every year.

Table 5.15

IRRIGATED AND RAINFED AREA UNDER MULBERRY CULTIVATION IN KARNATAKA, (1980-81 - 1993-94)

(Area in 000' hectares)

Year	Area Under Mulberry			
	Irrigated	Rainfed	Total	Percentage to India's Total
1980-81	51.01	65.18	116.19	68.35
1981-82	54.66	65.18	119.84	66.60
1982-83	57.89	65.18	123.07	62.52
1983-84	60.73	65.18	125.91	60.85
1984-85	64.13	63.97	128.10	59.63
1985-86	71.34	62.67	134.01	61.52
1986-87	78.58	62.19	140.77	61.27
1987-88	79.50	60.96	140.46	58.14
1988-89	80.32	60.97	141.29	52.71
1989-90	85.32	60.97	146.29	49.72
1990-91	89.89	60.97	150.86	48.18
1991-92	91.76	61.32	153.08	46.22
1992-93	103.44	54.50	157.94	46.07
1993-94	106.34	54.50	160.84	47.05

Source : Directorate of Sericulture, Bangalore.

The increase in area under irrigated mulberry after 1984-85 is mainly because of the constant failure of monsoons for the last several years, which has made it imperative for the farmers to grow the crop under irrigated conditions. On an average five crops can be obtained under irrigated conditions, and the yield per hectare is also twice more than that of rainfed mulberry. Table 5.15 gives the details of mulberry grown both under rainfed and irrigated conditions since 1980-81.

Table 5.15 reveals that the growth of the area under mulberry in Karnataka has been rather at a slow rate since 1980-81. The mulberry area under rainfed conditions shows a declining trend since 1984-85 and it reached to 54.50 thousand hectares by 1993-94. Simultaneously the area under irrigated conditions has gone up to reach 106.34 thousand hectares. Further, it is also observed that the contribution of this state to India's total has declined gradually from 68.35 per cent to 47.05 per cent (Graph 5.4). The regression coefficient of these trends, in the area under mulberry are analysed and presented in Table 5.16.

Table 5.16

REGRESSION COEFFICIENTS OF AREA UNDER MULBERRY IN KARNATAKA

Constant	Regression Coefficient	Growth Rate	R^2	t- value
4.7391	0.025	2.518	0.987	31.026

Source : computed from the data drawn from Table 5.15

The above table 5.16 shows that the coefficients of the area under mulberry cultivation in Karnataka are significant at 1 per cent level implying that the growth rates are significant. The coefficient of determination is 0.98 implying that there is a steady growth during the above period.

COCOON PRODUCTION

In Karnataka the production of reeling cocoons is steadily rising along with the rise in the area under mulberry cultivation. The yearwise quantity of reeling cocoons produced in the state and its share in India's total is given in Table 5.17.

From Table 5.17 it may be noted that there is a gradual increase in the quantity of cocoons produced between 1980-81 and 1993-94. When compared to India's total the state's share to it has gradually become reduced (Graph 5.5). The growth rate of cocoon production in the state is given in Table 5.18.

Table 5.17

REELING COCOONS PRODUCED IN KARNATAKA (1980-81 - 1993-94)

Year	Quantity of Cocoons Produced (000' tonnes)	Percentage to India's Total
1980-81	38.01	65.30
1981-82	33.52	60.71
1982-83	37.41	55.99
1983-84	40.13	56.30
1984-85	40.59	54.21
1985-86	42.67	55.62
1986-87	44.37	54.39
1987-88	47.22	54.57
1988-89	50.52	52.36
1989-90	57.72	52.27
1990-91	59.03	50.60
1991-92	51.97	48.50
1992-93	65.57	50.56
1993-94	70.21	57.00

Source : Directorate of Sericulture, Bangalore.

From Table 5.18 it is seen that the cocoon production is increasing at the growth rate of 5.27 per cent per year from 1980-81 to 1993-94. The coefficient of determination R^2(0.920) is significant at one per cent level indicating that the growth rate is significant.

Table 5.18

GROWTH RATE OF PRODUCTION OF REELING COCOONS IN KARNATAKA STATE

Constant	Regression Coefficient	Growth Rate	R^2	t- value
3.4730	0.051	5.265	0.920	11.743

Source : Computed from the data drawn from Table 5.17

RAW SILK PRODUCTION

Karnataka holds its supremacy by producing the largerer quantities of raw silk in the country. It accounts for nearly 54.05 per cent of India's total raw silk production. Its share was even more in 1980-81. Trends in raw silk production in Karnataka since 1980-81 is shown in Table 5.19.

Table 5.19

TRENDS IN RAW SILK PRODUCTION IN KARNATAKA (1980-81 - 1993-94)

Year	Raw Silk Production (000' Tonnes)	Percentage to India's Total
1980-81	2.88	62.75
1981-82	3.00	62.50
1982-83	3.20	61.42
1983-84	3.35	58.98
1984-85	4.06	58.84
1985-86	4.30	61.17
1986-87	4.67	59.11
1987-88	4.97	58.75
1988-89	5.32	54.96
1989-90	6.08	55.73
1990-91	6.21	54.05
1991-92	5.49	51.50
1992-93	7.92	56.10
1993-94	8.25	61.57

Source : Directorate of Sericulture, Bangalore.

As seen from the Table 5.19 the production of raw silk was only 2.88 thousand tonnes in 1980-81 and it gradually rose to 8.25 thousand tonnes by 1993-94 (Graph 5.6). The regression coefficient of raw silk production in the state is shown in Table 5.20.

Table 5.20

GROWTH RATE OF RAW SILK PRODUCTION IN KARNATAKA

Constant	Regression Coefficient	Growth Rate	R^2	t- value
0.9627	0.078	8.078	0.961	17.157

Source : Computed from the data drawn from Table 5.19

As seen from Table 5.20 in Karnataka the growth rate of raw silk production was 8.08 per cent which is significant at 1 per cent level.

DISTRICTWISE STATUS OF SERICULTURE

Mulberry cultivation has become one of the important commercial activities especially in the five traditional sericulture districts of Karnataka. The districtwise distribution of area under mulberry and cocoon production in Karnataka for the year 1993-94 is given in Table 5.21.

It is clear from Table 5.21 that the concentration of sericulture is mainly in the districts of Mysore, Kolar, Mandya, Bangalore (R) and Tumkur. Though the area under mulberry cultivation is greater in Mysore, i.e., 58,894 hectares, only 24 per cent of it, i.e., 13, 924 hectares, is under irrigated conditions and the remaining 76 per cent i.e., 44,970 hectares is under rainfed conditions. Where as in the Kolar district out of the total Mulberry area of 26,847 hectares an area of 25,624 hectares (95 per cent) is under irregated conditions. Therefore, it is evident from the table that the Kolar district ranks first in Karnataka in terms of area under irrigated mulberry, and also cocoon production. District wise intensity of area under mulberry is shown in Map 5.2.

Table 5.21

DISTRICT WISE STATUS OF SERICULTURE IN KARNATAKA (1993-94)

S.No.	District	Area under Mulberry (in Hectares)			Cocoon Production (in tonnes)
		Rainfed	Irrigated	Total	
1.	Bangalore (R)	1628	19103	20,731	14,504
2.	Bangalore (U)	26	2549	2,575	1,478
3.	Belgaum	--	1438	1,438	347
4.	Bellary	--	976	976	282
5.	Bidar	--	318	318	81
6.	Bijapur	--	993	993	215
7.	C.Mangalur	118	1125	1,243	214
8.	Chitradurga	65	3185	3,250	1,407
9.	D.Kannada	348	653	1,001	144
10.	Dharwad	--	1200	1,200	198
11.	Gulbarga	--	1266	1,266	246
12.	Hasan	2870	2764	5,634	1,352
13.	Kodagu	472	300	772	80
14.	Kolar	1223	25624	26,847	20,002
15.	Mandya	--	15267	15,267	9,776
16.	Mysore	44970	13924	58,894	14,151
17.	Raichur	--	702	702	134
18.	Shimoga	253	756	1,009	164
19.	Tumkur	2526	13428	15954	5,349
20.	U.Kannada	--	765	765	84
	Total	54499	10636	160883	70208

Source : Directorate of Sericulture, Bangalore.

MARKETING OF COCOONS

The cocoons produced are marketed in the existing government cocoon markets. The quantity of cocoons transacted in government cocoon markets is very high and the market fee collected is also very high. This shows that the accessibility of markets is good in Karnataka when compared to Andhra Pradesh. Adequate marketing facilities have made the sericulture industry well established in this state. The details of the cocoons transacted and the market fee collected from 1986-87 to 1991-92 are furnished in Table 5.22.

Table 5.22

TRANSACTIONS MADE IN GOVERNMENT COCOON MARKETS IN KARNATAKA STATE (1986-87 - 1991-92)

Year	Quantity of Cocoons Transacted (in Tonnes)	Total ValueRealised (Rupees in Lakhs)	Market Fee Collected (Rupees in Lakhs).
1986-87	39,945	17,059.59	343.7
1987-88	40,346	21,875.13	430.9
1988-89	44,382	26,658.80	509.8
1989-90	48,978	35,000.00	685.8
1990-91	57,460	38,110.00	741.4
1991-92	40,253	26,342.15	996.8

Source : Directorate of Sericulture, Bangalore.

From Table 5.22 it is seen that there are wide fluctuations in the quantity of cocoons transacted and the total value realised indicating wide fluctuations in the price of cocoons. The details of statistical analysis are given in Table 5.23.

Table 5.23

GROWTH RATE OF THE TRANSACTIONS MADE IN GOVERNMENT COCOON MARKETS IN KARNATAKA

Constant	Regression Coefficient	Growth Rate	R^2	t- value
10.5906	0.034	3.481	0.194	0.982

Source : Computed from the data drawn from Table 5.22

Table 5.23 shows that though the growth rate in the transactions made at the government cocoon markets is 3.481 per cent it is not statistically significant. The multiple correlation coefficient (R^2 0.194) is very low indicating that there are wide fluctuations in the quantity of cocoons transacted.

INFRASTRUCTURAL FACILITIES

As already mentioned earlier in this chapter, infrastructural facilities have a vital role to play in determining the pace of development of any industry. Here a picture of infrastructural facilities available in Karnataka for sericulture development is given in Table 5.24.

Table 5.24

INFRASTRUCTURAL FACILITIES IN KARNATAKA (1991)

Infrastructure Available	Number
Government Silk Farms	95
Government Grainages	81
Grainages of CSB	8
Technical Service Centres	150
Extension-cum-chawkie rearing centres	20
Tribal extension centres	10
Government Chawkie rearing Centres	1238
Private chawki rearing centers	468
Mounting halls	2
Cocoon markets	59
Spun Silk Mills	1
Sericulturists-cum-farmers co-operatives	109
Silk exchanges	6

Source : Directorate of Sericulture, Bangalore

As sericulture in Karnataka is deep-rooted for centuries and enjoys the prestige of a traditional commercial crop, infrastructural facilities are vastly developed. In 1991 there were 95 government chawkie rearing centres and 59 cocoon markets working in all over the state.

SERICULTURE PROGRAMMES AND PROJECTS IN KARNATAKA

An uninterrupted flow of funds from both the central and state governments under different schemes and projects has made sericulture to achieve its present status. A brief description of the various schemes and projects undertaken by the Karnataka state government are given in the following pages.

NATIONAL SERICULTURE PROJECT (KARNATAKA SERICULTURE PROJECT - PHASE-II)

Karnataka successfully implemented the Karnataka Sericulture Project - Phase-I, assisted by the World Bank during the period from April 1980 to

September 1988. The project cost was Rs.101.80 cores.

The Department of Sericulture of the state prepared the second phase for the Sericulture Project with an outlay of Rs.239.43 cores to be proposed to the World Bank with greater emphasis on seed production, strengthening of extension, cocoon markets, training and Post-cocoon processing. However, in the final stage the second phase of the sericulture project of Karnataka was incorporated with the National Sericulture Project. For Karnataka the size of the approved project is Rs.118.3 cores.

The Objectives of the Project are:

a) to increase silk production by about 3346 tonnes of which 500 tonnes will be bivoltine silk;

b) to bring an additional area of 15,000 hectares under mulberry cultivation;

c) to increase the production of cocoons per hectare from 625 Kgs. to 750 Kgs. under irrigated mulberry and from 200 Kgs. to 230 Kgs. under rainfed mulberry;

d) to increase the average production of cocoons per 100 layings from 25 Kgs. to 30 Kgs. under irrigated mulberry and from 20 Kgs. to 23 Kgs. under rainfed mulberry;

e) to reduce the renditta from 10 to 8.5 and from 12 to 10.5 under irrigated and rainfed conditions respectively;

f) to generate incremental employment to about 2.71 lakh persons.

During the year 1990-91, as against the targetted area of 2,650 hectares, 3,500 hectares were brought under cultivation even by the end of October. The cumulative achievement was 8,500 hectares as against the target of 4,550 hectares accounting for 186.8 per cent increase. A total of 2,219 farmers were trained in sericulture activities, of whom 1,368 were women, accounting for 62 per cent. The number of persons trained from the commencement of the project was 4,537 of whom women were 2,972 accounting for 65 per cent. The project also supports each grainage with working capital of Rs.12 lakhs. To operate the funds, the government has been requested to accord permission to open Personnel Deposit Account.

But the proposal is still pending before the Finance Department. In addition to these, the scheme, provides adequate support to the sericulturists, to rear bivoltine silkworms, and to promote reeling activity.[17]

DUTCH AIDED PROJECT

Sericulturists are being helped to improve their rearing facilities under this scheme. The outlay of this project is Rs.37.241 million. The project envisages construction of 238 living-cum-rearing sheds and 500 rearing houses for the beneficiaries and 122 chawkie rearing buildings. Individual beneficiaries are selected for according sanction of loan for the construction of sheds. Preference has been given to small and marginal Scheduled Caste and also for Scheduled Tribe sericulturists. Interest charges are 5.5 per cent and the loan is repayable in 19 years in equal amounts after a moratorium of two years for principal. The Small, Marginal, Scheduled Caste and Scheduled Tribe sericulturists are exempted from payment of registration fee and stamp duty.[18]

SPECIAL COMPONENT PLAN

Sericulture, which plays an important role in alleviating unemployment and rural poverty in general, also plays a major role in the upliftment of farmers belonging to the Scheduled Castes and Scheduled Tribes. Providing training, creating of irrigation facilities, supply of saplings, supply of rearing and reeling equipments and such programmes are under implementation under this Special Component Plan. Incentives worth of Rs.300/- to Rs.500/- towards the cost of mulberry cuttings and sixty per cent subsidy are given to the financial institutions in the name of the rearers for rearing building worth Rs.20,000/-. The Scheduled Caste sericulturists who have taken up mulberry plantation newly will be given rearing equipment whose worth ranges from Rs.1000/- to Rs.3000/- per 0.40 hectare per person free of cost. Package benefits like nylon nets, inputs etc., worth ranging from Rs.400/- to Rs.2000/- is given free of cost per 0.40 hectare.

During 1989-90 Rs.29.44 lakhs were spent as against an outlay of Rs.36.57 lakhs. And a sum of Rs.27.90 lakhs was allotted under Special Central Assistance as against Rs.28.76 lakhs was spent.[19] During 1990-91 the expenditure incurred under this plan was Rs.8.98 lakhs.

TRIBAL SUB PLAN

The Tribal sub-plan takes an integrated view of the problems of the tribal people. In the given district under the Tribal Sub-plan incentives worth of Rs.3000/- per 0.40 hectare per person towards the cost of mulberry cuttings, inputs and rearing equipments are given. Package benefits worth ranginhg from Rs.2000/- to 4000/- like nylon nets, inputs etc., are given free of cost per 0.40 hectare per person.

During 1989-90 an outlay of Rs.10.00 lakhs was provided under the state sector, of which Rs.9.37 lakhs was spent. During 1990-91, Rs.8.11 lakhs was spent (out of an outlay of Rs.9.50 lakhs). Under the Central Sector, during 1990-91 an amount of Rs.6.20 lakhs was provided and Rs.9.11 lakhs were spent. In the Eight Five Year Plan, the outlay set apart for the Special Component Plan is Rs.10 cores and for the Tribal Plan the outlay is Rs.0.40 cores.[20]

Apart from the programmes mentioned above for the development of Sericulture, there are the Western Ghats Scheme, the Karnataka Twenty Point Programme and the Karnataka Development Programme which are also under implementation in the State

CONCLUSION

The Karnataka State has been regarded as the Silk Bowl of India and accounts for more than fifty per cent of India's total raw silk production. Andhra Pradesh follows it in this regard. The general climatic conditions in the two states are quite favourable, for the growth of sericulture. Sericulture though new to Andhra Pradesh, has gained remarkable momentum because of the sincere efforts of the state government. However in Karnataka sericulture, apart from receiving governmental support, has the status of one of the traditional occupations and is closely inter-woven with the living styles of the people. There has been a tremendous growth in the area under mulberry cultivation since 1980-81 in both the states. In Andhra Pradesh the area under mulberry went up from 18,900 hectares in 1980-81 to 90,774 hectares in 1993-94, whereas in Karnataka it rose from 1,16,190 hectares in 1980-81 to 1,60,840 hectares in 1993-94. The growth rate of the area under mulberry (11.609 per cent) is higher in Andhra Pradesh though it is a new enterprise to this state. But Karnataka has a larger area under mulberry cultivation than Andhra Pradesh. Further, in the production of reeling cocoons and raw silk, Karnataka has retained its supremacy. It contributes more than 50 per cent of the total reeling cocoons and raw silk produced in the country. Owing to the well-established marketing facilities in Karnataka the transactions of cocoons are much better in Karnataka than in Andhra Pradesh. Hence, there is the need to enhance the marketing facilities in Andhra Pradesh to raise the revenue through sericulture. As the policy of the government of India towards sericulture has been favourable, Karnataka and Andhra pradesh have been undertaking a number of developmental programmes and projects. Of them the recent National Sericulture Project with World Bank Assistance needs special mention. Apart from this, in both states there are many other projects under implementation contributing to the development of sericulture. India can achieve the first rank in the production of raw silk before long if the present tempo of development is sustained and continued in these two states.

Notes:

1. Andhra pradesh State Gazetteer - Government of India.
2. Hand Book of Statistics 1990-91, Directorate of Economics and Stitistics, Government of A.P., Hyderabad.
3. Government of India - Drought Prone Areas Programme, Ministry of Agriculture and Irrigation (Department of Rural Development), New Delhi, 1978.
4. Hand Book of Statistics 1990-91 - Directorate of Economics and Statistics, Op cit.
5. New Horizons for Sericulture in Andhra Pradesh, Department of Textiles and Handlooms, 1976, p.6.
6. Smt. Asha Murthy, "Sericulture in Andhra Pradesh", Souvenir on International Congress on Tropical Sericulture Practices, February, 1988, Op.cit., p.89.
7. "Annual Administration Report - 1991-92", Directorate of Sericulture, Hyderabad, p.1.
8. Ibid., p.3.
9. Md. Muneer Pash, "Swiss aid to the Mulberry Sericulture Development Project in Andhra Pradesh and Tamil Nadu", Indian Silk, Vol.XXVII, No.3, July 1988, p.28.
10. "Annual Administration Report 1991-92" - Department of Sericulture, Hyderabad, p.8.
11. Ibid., p.4.
12. "Annual Administration Report 1990-91", Department of Sericulture, op.cit., p.5
13. Vasant Desai, Rural Development Vol.II, Programmes and Strategies, Himalaya Publishing House, 1988, p.148.
14. "Annual Administration Report 1990-91", Op.cit., p.5
15. Ibid. p.5
16. Directorate of Economics and Statistics, "District Socio-Economic Indicators, 1993", p.5.
17. "National Sericulture Project-Notes for Review", Directorate of Sericulture Bangalore, 1992, pp 1-9.
18. "Annual Report 1982-83", Directorate of Sericulture, Bangalore, p.8.
19. "Annual Administrative Report 1989-90", Directorate of Sericulture, Bangalore, p.10.
20. "Annual Administrative Report 1990-91", Directorate of Sericulture, Bangalore, p.11.

CHAPTER 6

Prospects of Sericulture in the Rural Areas An Emperical Analysis

As already mentioned in the earlier chapters, sericulture has been making rapid strides of progress in Andhra Pradesh and Karnataka particularly in the Anantapur and Kolar districts in their respective states. These two districts have the unique distinction of producing mulberry silk and also have sericulture deeply rooted in almost all of their blocks. In these districts sericulture is practiced under irrigated conditions. Inspite of severe drought and low rainfall conditions, mulberry cultivation in the Anantapur district has increased enormously during the last 15 years. Similarly the Kolar district which is also drought affected and suffers from variations in rainfall from year to year has registered remarkable progress. As mulberry is drought resistant, sericulture has become the most promising activity to the farmers of these districts. The success of sericulture in Anantapur and Kolar is largely attributable to the efforts of the respective Government Departments of Sericulture.

In this chapter an attempt is made to anayse in detail sericulture development in the Anantapur and Kolar districts. It is felt that a brief discussion of the economy, growth and present status of sericulture in these two districts is essential for the purpose of full length review. The present chapter is divided into three sections. In the first section, the progress and status of sericulture in Anantapur and Kolar districts are discussed. In the second section the socio-economic status of the respondents is presented. In the third section, the economics of sericulture and the prospects of sericulture are examined with empirical analysis.

SECTION - I

DEVELOPMENT OF SERICULTURE IN ANANTAPUR DISTRICT

The Anantapur district was formed in the year 1882 separating it from the then Bellary district. This district lies between 13⁰ 14' - 15⁰ 15' North latitudes and 76⁰ 51' - 78⁰ 30' East longitudes. It is the southern most district of the Rayalaseema Region of Andhra Pradesh. Anantapur district is bounded the by Cuddapah, Kurnool, and Chittoor districts on the east, north, south respectively and by the Karnataka state on the west[1]. The district has been divided into 63 revenue mandals under three revenue divisions viz., Anantapur, Dharmavaram and Penukonda. (Map 6.1).

As per the 1991 census, the total population of the district is 31,83,814 of which 24,35,761 (77 per cent) live in rural areas. The density of population is 166 per sq. km. The working force forms 43 per cent of which 39 per cent is in the agricultural sector[2].

The geographical conditions of the district which is in the middle of the peninsula, render it as the driest part of the state and hence, agricultural conditions are more often than not precarious. Monsoons evade this part due to its unfortunate location. Being far away from the east coast it does not enjoy the full benefits of the North-east monsoons and being cut-off by the Western Ghats, the South-west monsoons are also prevented from penetrating into this district and quenching the thirst of its parched soils. Therefore, this district is deprived of the benefits of both the monsoons, and is subject to recurrent droughts and bad seasons. The normal rainfall in the district is 544 mm, the lowest in comparison with that of the other Rayalaseema districts and other parts of Andhra Pradesh. However the district has a fairly good elevation, which provides the district with a tolerable climate, throughout the year.

The soils in the Anantapur district are predominantly red (76.5 per cent) and there are about 23.5 per cent of black soils.[3] Agriculture is the main occupation in the district. The total geographical area of the district is 19.13 lakh hectares, of which the normal cultivated area accounts for 9.66 lakh hectares. The district occupies the second lowest position in Andhra Pradesh with regard to surface irrigation facilities and hence the farmers have to depend more and more on groundwater resources which are being exploited through dug wells and borewells[4]. The main crops grown in the district are Paddy, Wheat, Jowar, Ragi, Pulses, Groundnut, Mulberry and other millets. Paddy and Jowar are the major food crops whereas Groundnut and Mulberry are grown as important commercial crops . The distribution of area under food and non-food crops from 1980-81 to 1991-92 is shown in Table 6.1.

It is clear form Table 6.1 that the area under non-food crops increased form 44 to 79 per cent over a period of 11 years.. This shows the rising interest of the farmers in non-food crops which have commercial value. There has been a slow transformation from traditional crops to non-traditional crops in the district. Although a number of commercial crops are under cultivation in the district, mulberry seems to dominate over many others, both in terms of area and returns. Sericulture has thus attracted a large number of farmers and it has occupied a significant place in the agricultural economy of the district. A brief description of the development of sericulture in it is given below.

As noted already, Anantapur occupies the first place in sericulture development in the state of Andhra Pradesh as mulberry is fully concentrated in this district. The district is widely known for its "Dharmavaram Handloom Silk Sarees". The major factors responsible for the concentration of sericulture in the district or its favourable climatic conditions conducive to silkworm rearing and cocoon production. The need for diversification in cultivation to take advantage of the facilities offered by the Government in the form of land development, water and power resources, technical assistance, marketing facilities and high returns has prompted several agriculturists in the district to switch over to mulberry cultivation and silkworm rearing. Another important and influential factor for the enormous growth of sericulture in this district is the availability of finances under the Five Year Plan programmmes and the Drought Prone Areas programme.

Talamarla, a village of the erstwhile Kadiri taluk, first appeared in the year 1923-24 on the sericulture map of Anantapur. In 1926, Penukonda came under mulberry cultivation and an Assistant Demonstrator was appointed there to promote sericulture. Demonstration of silkworm rearing and mulberry plantation were taken up by him and the farmers were encouraged to adopt sericulture as a subsidiary crop.[5] But cultivation of mulberry as a subsidiary crop did not gain the farmers' support. Again in 1946, Sri Venkatappa, a resident of Veebhuthi palli, now belonging to the Lepakshi mandal started cultivating mulberry in an area of 2.00 hectares. However cultivation of mulberry did not make much head way for another two and a half decades due to lack of knowledge and Government assistance. It gained momentum only after the implementation of the Drought Prone Areas Programme in the district in 1975.

Sericulture was in an embryonic stage in the distrct prior to the commencement of this programme. During the implementation of the Drought Prone Areas Programme, particular emphasis was laid on the development of sericulture as it was identified as an effective tool for rural development, as it had a number of plus points in its favour.

Table 6.1

AREA UNDER FOOD AND NON-FOOD CROPS IN ANANTAPUR DISTRICT, (1980-81 - 1991-92)

Year	Area under food crops	Area under non-food crops	Total area
1980-81	4.86 (56)	3.82 (44)	8.68 (100)
1981-82	5.50 (56)	4.40 (44)	9.90 (100)
1982-83	4.76 (52)	4.38 (48)	9.14 (100)
1983-84	4.58 (48)	4.99 (52)	9.57 (100)
1984-85	3.57 (41)	5.12 (59)	8.69 (100)
1985-86	2.94 (36)	5.25 (64)	8.19 (100)
1986-87	3.14 (40)	4.67 (60)	7.81 (100)
1987-88	3.12 (35)	5.88 (65)	9.00 (100)
1988-89	--	--	--
1989-90	2.69 (26)	7.74 (74)	10.43 (100)
1990-91	2.01 (20)	7.93 (80)	9.94 (100)
1991-92	2.09 (21)	7.86 (79)	9.95 (100)

Source : Directorate of Economics and Statistics, Hyderabad
Note : Figures in parentheses indicate percentages

There were hardly 3,560 mulberry growers in the district at that time. The industry had to pass through teething problems with regard to the supply of mulberry cuttings and disease-free layings to the farmers. The price of cocoons was low and could not attract farmers initially. Marketing facilities for mulberry cocoons were also lacking. But the major constraint was the meager budget that restricted the vertical growth of the industry.

Slowly, with the gradual rise in the price of cocoons and incentives provided to farmers for plantation, construction of rearing sheds, sinking of new wells and improvement in the supply of highly productive cross-breeds, establishment of cocoon markets and adequate budget under the Drought Prone Areas Programme, all these together created ideal conditions for the astonishing progress of sericulture in the district. Now sericulture is being practiced in almost all the 63 mandals of the district of which the mandals of seven erstwhile taluks, Hindupur, Kadiri, Madakasira, Kothacheruvu, Dharmavaram and Penukonda are well-known for sericulture activity and account for nearly 30 per cent of the total area under mulberry cultivation.

With the encouragement given by the state government sericulture in Anantapur has been progressing to reach commendable heights. Mulberry cultivation and silkworm rearing has become a boon and transformed the life styles of several farmers, especially the small and marginal farmers. By 1991-92 there were about 50,702 sericulturists in the district. Categorywise distribution of these farmers is shown in Table 6.2.

Table 6.2

CATEGORYWISE DISTRIBUTION OF SERICULTURISTS IN ANANTAPUR DISTRICT (1991-92)

Category of the farmer	Number	Percentage to total
Marginal	16,006	31.57
Small	20,589	40.61
Big	14,107	27.82
Total	50,702	100.00

Source : Deputy Director, Department of Sericulture, Anantapur

From Table 6.2 it is evident that small and marginal farmers constitute nearly 72.18 per cent and the remaining 27.82 per cent are big farmers. Castewise distribution of the Sericulturists is given in Table 6.3.

Table 6.3

CASTEWISE DISTRIBUTION OF SERICULTURISTS IN ANANTAPUR DISTRICT (1991-92)

Caste	Number of Sericulturists	Percentage to the total
SC	3,415	6.74
ST	1,767	3.48
BC	19,767	38.99
OC	25,753	50.79
Total	50,702	100.00

Source : Deputy Director, Department of Sericulture, Anantapur

Table 6.3 shows that a majority of the sericulturists belong to the other castes category with 50.79 per cent, followed by the backward castes with

Graph 6.1

Area Under Mulberry and Cocoon Production In Anantapur District

(1980-81 - 1993-94)

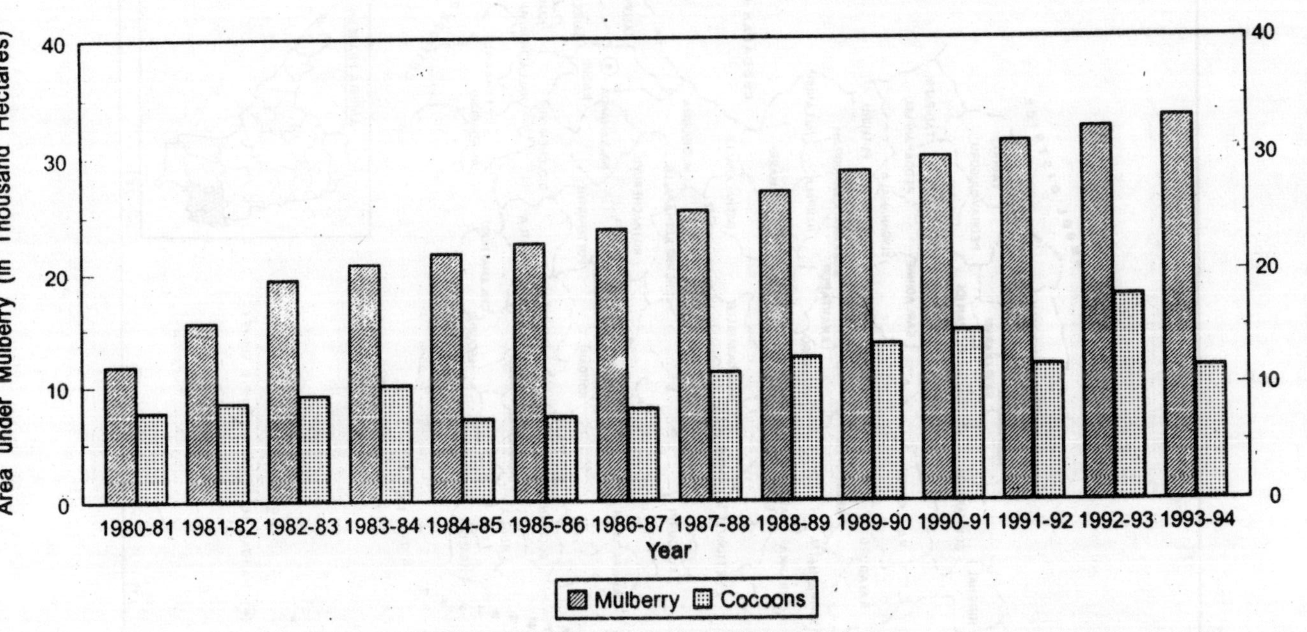

Map 6.1
Anantapur District

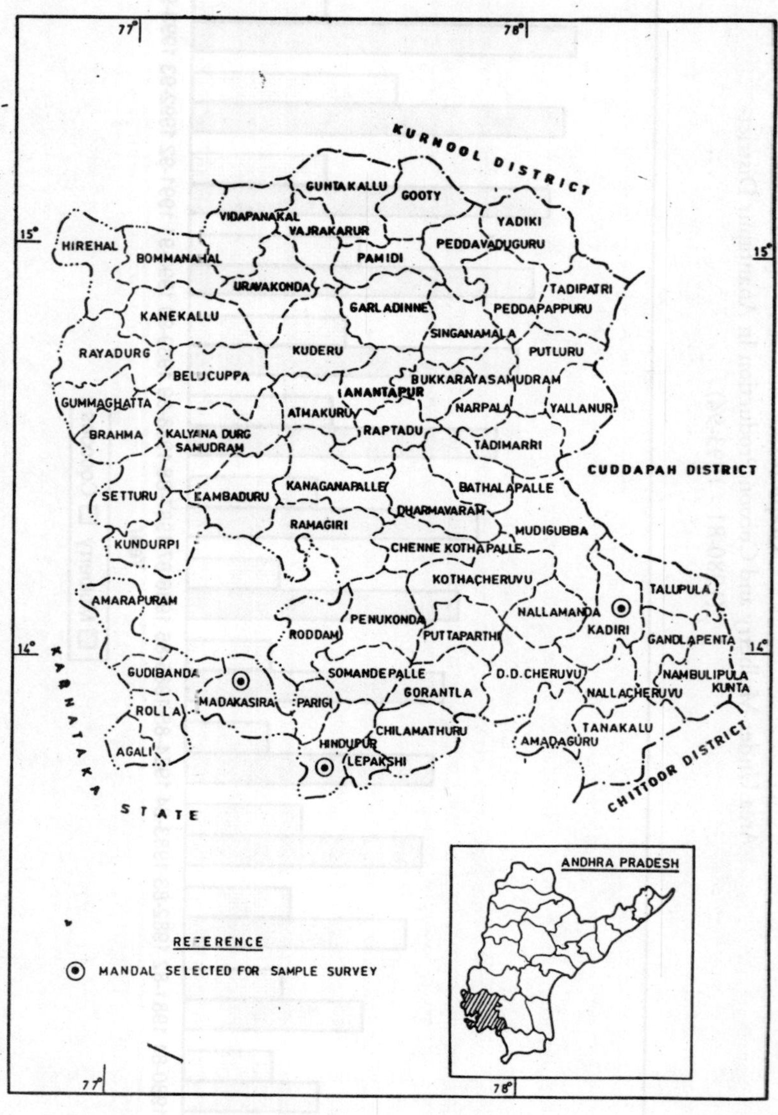

38.99 Per cent. Scheduled Castes and Scheduled Tribes constitute 6.74 and 3.48 per cent respectively.

With the increase in the number of sericulture growers year after year the area under mulberry as well as cocoon production has gone up steeply. Yearwise growth from 1980-81 to 1993-94 is shown in Table 6.4.

Table 6.4

AREA UNDER MULBERRY AND COCOON PRODUCTION IN ANANTAPUR DISTRICT (1980-81 - 1993-94)

Year	Area under mulberry (in hectares)	Cocoons Produced (in tonnes)
1980-81	11,782	7,721
1981-82	15,559	8,512
1982-83	19,416	9,211
1983-84	20,768	10,109
1984-85	21,663	7,060
1985-86	22,510	7,281
1986-87	23,732	7,950
1987-88	25,311	11,132
1988-89	26,885	12,400
1989-90	28,589	13,561
1990-91	29,861	14,737
1991-92	31,167	11,713
1992-93	32,340	17,901
1993-94	33,220	11,470

Source : Directorate of Sericulture, Hyderabad

It is seen from Table 6.4 that the area under mulberry which was only 11,782 in 1980-81 went up to 33,220 hectares in 1993-94 (Graph 6.1). This reveals the increasing attraction of mulberry cultivation has had for farmers. Similarly, the cocoons produced also rose from 7,721 Tonnes in

1980-81 to 11,470 Tonnes in 1993-94 (Graph 6.1). It may be observed that there are wide fluctuations in the production of reeling cocoons. This may have been mainly due to lack of timely supply of quality layings from the government grainages, adverse climatic conditions, use of old and outdated rearing appliances, and lack of timely technical assistance.

The growth rates in the area under mulberry hecterage and cocoon production are statistically tested and the results are presented in Table 6.5

Table 6.5

GROWTH RATES OF AREA UNDER MULBERRY AND COCOON PRODUCTION

Category	Constant	Regression Coefficient	Growth Rate	R^2	t- value
Area under Mulberry	2.6617	0.067	6.889	0.895	10.120
Cocoon Production	1.9542	0.051	5.260	0.575	4.028

Source : computed from the data drawn from Table 6.4.

It is evident from Table 6.5 that the area under mulberry and cocoon production in the district has been increasing at the rate of 6.889 per cent and 5.260 per cent respectively per year, between 1980-81 and 1993-94. The coefficient of determination, R^2, (0.895) and 't' (10.120) are significant implying that the growth rate of the area under mulberry is significant at 1 per cent level. Similarly, the growth rate of cocoon production is also significant at 1 per cent level.

The rise in the production of reeling cocoons has simultaneously resulted in the production of huge quantities of raw silk in this district. The raw silk produced in the government sector from the year 1987-88 to 1993-94 is presented in Table 6.6.

It is evident from the Table 6.6 that the production of raw silk in Anantapur district has been showing an upward shift since 1987-88. But there was a decline in raw silk production during the year 1993-94, may be due to the intensity of diseases affecting the silkworms in the last stages of their rearing, and severe heat during summer. Statistical analysis of the growth rate of raw silk production is presented in Table 6.7.

It is evident from Table 6.7 that the growth rate of mulberry raw silk is high at 11.32 per cent per year. The coefficient of determination (R_2)0.739 and 't' value 3.764 are significant at 5 per cent level implying that there is a significant growth of mulberry raw silk production.

Table 6.6

RAWSILK PRODUCTION IN ANANTAPUR DISTRICT
(1987-88 TO 1993-94)

(in '000 Tonnes)

Year	Raw silk Production
1987-88	0.19
1988-89	0.21
1989-90	0.24
1990-91	0.34
1991-92	0.34
1992-93	0.38
1993-94	0.31

Source : Directorate of Sericulture Hyderabad.

Table 6.7

GROWTH RATE OF RAWSILK PRODUCTION IN ANANTAPUR DISTRICT

Constant	Regression Coefficient	Growth Rate	R^2	t- value
-1.7069	0.107	11.322	0.739	3.764

Source : computed from the data drawn from Table 6.6.

As the hectarage under mulberry increased year after year, the demand for cross-bread disease-free layings also increased simultaneously. Table 6.8 gives the picture of the disease-free layings produced in the government grainages in the district.

Table 6.8 shows that the production of disease-free layings has registered an upward trend during the period 1985 to 1992 except in 1990-91.

Regarding marketing, previously the sericulturists were transporting their cocoons to the Karnataka cocoon markets. Later on the government of Andhra Pradesh established cocoon markets at Hindupur, Kadiri and Dharmavaram under the provision of Andhra Pradesh Silkworm Seed Cocoon Control Act 1956. The cocoon markets started functioning from April 1983. Yearwise quantity of cocoons transacted, value of cocoons transacted and market fee collected since 1986-87 are presented in Table 6.9.

Table 6.8

PRODUCTION OF DISEASE FREE LAYINGS (DFLS) IN GOVERNMENT GRAINAGES IN ANANTAPUR DISTRICT (1985-86 to 1991-92)

Year	Dfls in Lakh Nos.
1985-86	69.77
1986-87	73.16
1987-88	95.02
1988-89	105.22
1989-90	109.12
1990-91	108.51
1991-92	165.40

Source : Deputy Director, Department of Sericulture, Anantapur

Table 6.9 makes it clear that the cocoon markets are fetching huge revenue to the government and extending service to sericulturists. There have been fluctuations in the quantity of cocoons transacted and the total value realised because of market imperfections and lack of knowledge about the prices prevailing. However, since 1986-87 the quantity of cocoons transacted, the amount realised and the market fee collected have been showing an unmistakable increasing trend.

The vast growth in the area under mulberry, disease-free layings and production of reeling cocoons is mainly due to the enhanced infrastructural facilities made available. Since the inception of the DPAP, an amount of Rs. 6.40 crores flowed into the district between 1976-77 and 1988-89 for the development of sericulture. With such enormous funds flowing in there was, considerable increase in the infrastructure facilities which helped to build up the area under mulberry and cocoon production. The infrastructural facilities available during 1991 are shown in Table 6.10.

The vast growth in the area under mulberry, disease-free layings and production of reeling cocoons is mainly due to the enhanced infrastructural facilities made available. Since the inception of the DPAP, an amount of Rs. 6.40 crores flowed into the district between 1976-77 and 1988-89 for

Map 6.2
Kolar District

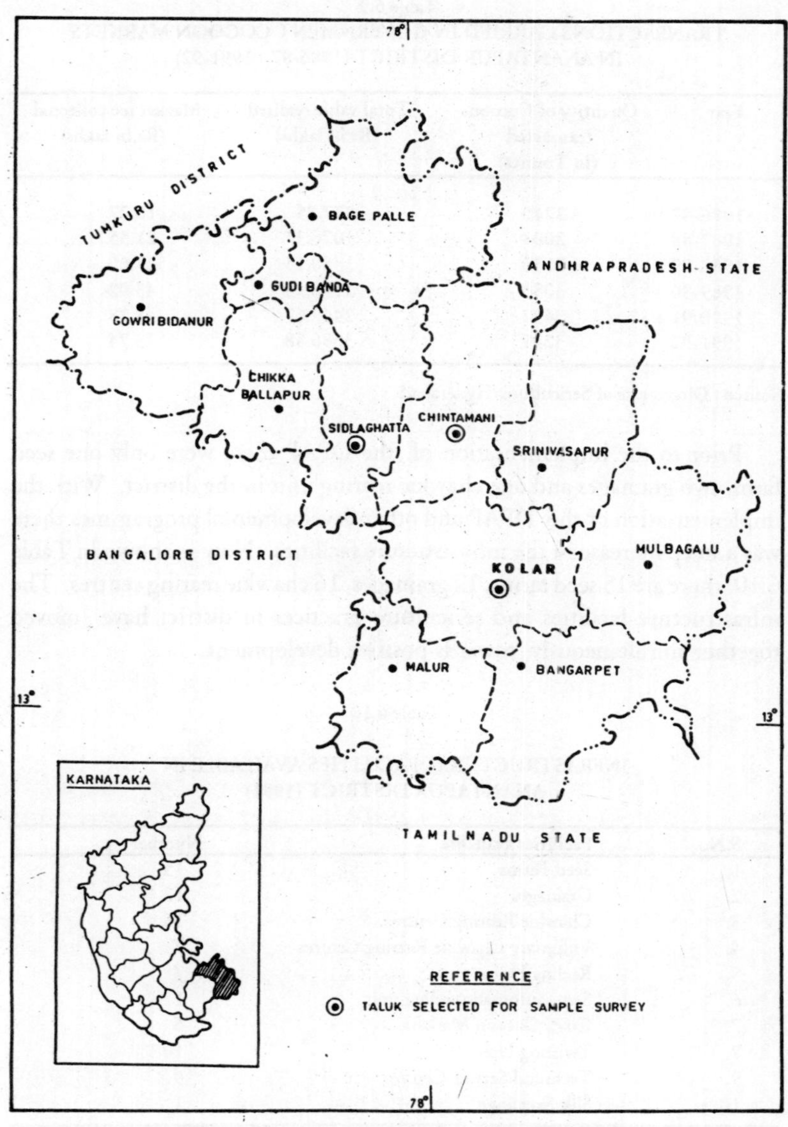

the development of sericulture. With such enormous funds flowing in there was, considerable increase in the infrastructure facilities which helped to build up the area under mulberry and cocoon production. The infrastructural facilities available during 1991 are shown in Table 6.10.

Table 6.9

TRANSACTIONS CARRIED IN GOVERNMENT COCOON MARKETS IN ANANTAPUR DISTRICT (1986-87 - 1991-92)

Year	Quantity of Cocoons transacted (in Tonnes)	Total value realised (Rs.in lakhs)	Market fee collected (Rs.in lakhs)
1986-87	2240	937.85	18.77
1987-88	2008	1076.37	21.55
1988-89	2130	1380.15	27.60
1989-90	3050	2176.13	43.02
1990-91	3492	2449.67	48.78
1991-92	3231	3686.58	73.74

Source : Directorate of Sericulture, Hyderabad.

Prior to the implementation of the DPAP there were only one seed farm, two grainages and one chawkie rearing unit in the district. With the implementation of the DPAP and other developmental programmes there was a steep increase in the infrastructure facilities. Now as shown in Table 6.10, there are 15 seed farms, 12 grainages, 16 chawkie rearing centres. The infrastructure facilities and sericulture practices in district have moved together simultaneously towards positive development.

Table 6.10

INFRASTRUCTURAL FACILITIES AVAILABLE IN ANANTAPUR DISTRICT (1991)

S.No.	Facilities available	Number
1.	Seed Farms	15
2.	Grainages	12
3.	Chawkie Rearing Centres	16
4.	Voluntary Chawkie Rearing Centres	7
5.	Reeling Units	7
6.	Semi-automatic reeling unit	1
7.	Govt. Cocoon Markets	6
8.	Twisting Units	10
9.	Technical Service Centres	9
10.	Silk Exchange	1

Source : Office of the Deputy Director, Sericulture, Anantapur.

Thus the district owes a lot to the Drought Prone Areas Programme through which enormous funds were flown into it for the development of sericulture. Now in addition to Plan funds, from various other schemes funds were flown into the district. They are the National Sericulture Project, the Indo-Swiss Programme, the Integrated Rural Development Programme and the Special Component Plan.

The Government of Andhra Pradesh has been giving several incentives to those who take up sericulture. To the small and marginal farmers, subsidy at the rate of 25 per cent is being given under the IRDP on the unit cost of Rs. 19,800/-. Fifty per cent subsidy is being given for supply of uzyfly nylon nets. Rs. 5/-per Kilogram is being given on production of bivoltine cocoons. Premium for crop insurance is being subsidised by fifty per cent for local race seed rearers. With all these encouragements Sericulture in Anantapur has been making commedable progress.

Now a brief description of development of the sericulture in the Kolar district of Karnataka which resembles Anantapur in many ways, may be attempted.

DEVELOPMENT OF SERICULTURE IN KOLAR DISTRICT

The Kolar district is situated in the southern region of Karnataka. It lies between 12° 46' - 13° 58' North latitudes and 77° 21' - 78° 35' East longitudes. It is bounded by the districts of Bangalore and Tumkur of Karnataka on the west, by the Anantapur and Chittoor districts of Andhra Pradesh on the north and on the east respectively, and by the North Arcot and Dharmapuri districts of Tamil Nadu on the South. This district is known for its gold mines. The gold-bearing lands of the Kolar Gold Fields lie in a narrow band of rocks of Dharwar series for about 80 km length from Sreenivasapur in the north to within a few Kilometers of Krishnagiri in Tamil Naidu.[6]

The district has been divided into two Revenue Sub-Divisions called Chikbalapur and Kolar, with six and five Taluks respectively under each sub-division (Map 6.2). As per the 1991 census the district has a population of 22,11,300; out of which 17,03,100 (77 per cent) live in rural areas. The density of population is 232 persons per sq.km. The population has recorded a growth of 16.05 per cent during 1981-91.[7]

There are no perennial rivers in the district. Most of the rivers are small and carry water only during the rainy season. Three important rivers, the

Graph 6.2
Area Under MUlberry and Cocoon Production in Kolar District
(1980-81 - -1993-94)

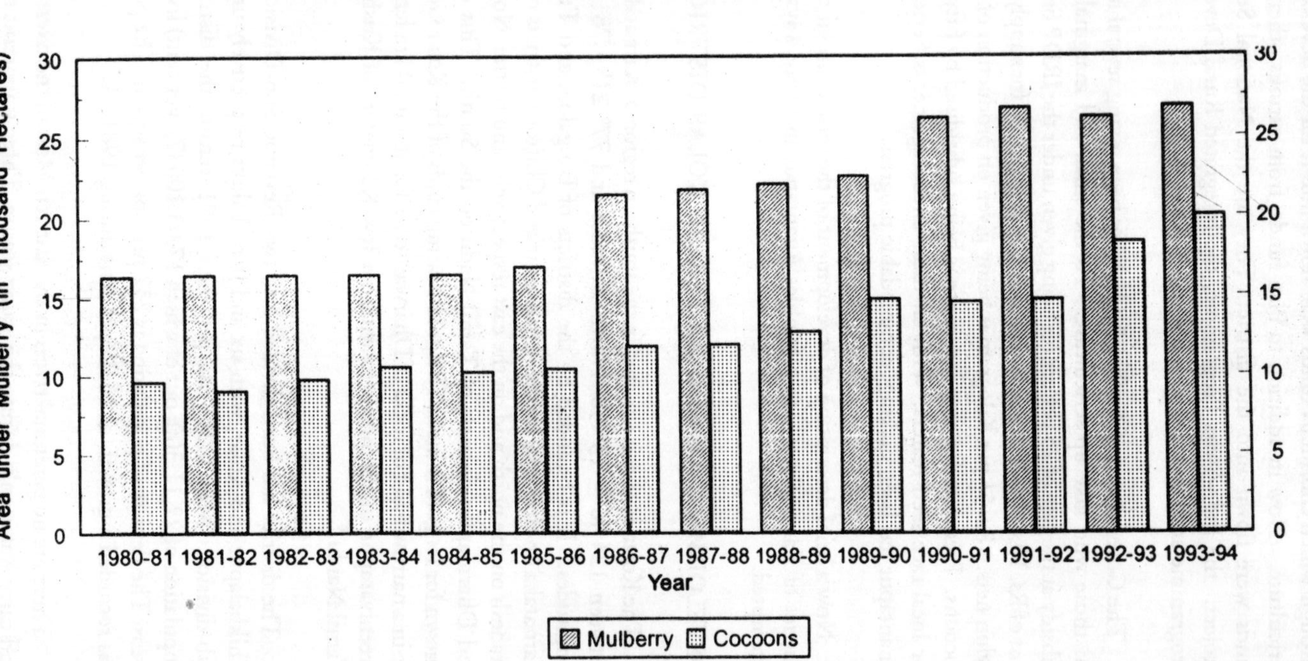

Palar, the North Pinakini and the South Pinakini and several of their tributaries drain the district. With these rivers and streams flowing in several directions and a large number of tanks and wells situated amidst hilly surroundings one would expect to find the district full of green vegetation. But the rainfall being scanty and the rivers and streams being dry for the better part of the year, the area is, for the most part, devoid of vegetation, and drought conditions are very common. However, the district enjoys an agreeable climate. The annual normal rainfall of the district is 733.9mm of which the contribution of South-west monsoon is 69 per cent. But failures of rains are frequent in the district, and therefore, droughts are common. Therefore this district is listed among the drought prone districts of Karnataka.[8]

Agriculture is the main occupation for the people in this district. Out of 8.22 lakh hectares nearly 3.87 lakh hectares (46 per cent) are under cultivation. Of the total cultivated land about 24 per cent is under irrigation and the remaining 76 per cent is under rainfed conditions. Nearly 72 per cent of the area under irrigation is through groundwater resources[9]. Thus groundwater resources have an importance in the development of the district.

The staple food crops grown in Kolar are Paddy and Ragi. Jowar, Bajra and Wheat are also grown in some parts of the district. Groundnut and mulberry are two important commercial crops grown in almost all parts of the district . Table 6.11 gives an idea about the area under food and non-food crops in the district.

It is seen from Table 6.11 that the area under food and non-food crops is gradually increasing, and the area under food crops is larger than that of non-food crops. But, the area under non-food crops is also slowly increasing. Thus non-food crops of commercial importance are gaining the attention of farmers. Among the various commercial crops mulberry is age old and a traditional crop, and has been playing a vital role in the agricultural economy of the district.

Today, the Kolar district stands first in the state in sericulture activities. Sericulture in this district has a long history dating back to the earlier part of the 19th century. During the Second World War the Silk industry in Kolar received special attention as pure silk was in great demand as it was the only fibre which was used for making parachutes.[10] The increased demand led to increased production in the filatures at Sidlaghatta of the district. As a result cocoon harvesting gained momentum in this district.

During the post-war period, the sericulture industry in the entire state faced a slump in filature production. The filatures were faced with financial difficulties. So the Government took over all the filatures except the one at Sidlaghatta. But the unit at Sidlaghatta could not be continued for long and it was closed down. After the reorganisation of states, the Government of the erstwhile Mysore state took over the filatures and established a number of grainages. The Kolar district also benefitted with this establishment of filatures and grainages under government control.

Table 6.11
AREA UNDER FOOD AND NON-FOOD CROPS IN
KOLAR DISTRICT (1980-81 - 1991-92)

Year	Food Crops	Non-food Crops	Total
1980-81	0.42	0.17	0.59
	(71)	(29)	(100)
1981-82	0.63	0.19	0.82
	(77)	(23)	(100)
1982-83	0.58	0.19	0.77
	(75)	(25)	(100)
1983-84	0.64	0.20	0.84
	(76)	(24)	(100)
1984-85	0.78	0.25	1.03
	(76)	(24)	(100)
1985-86	0.68	0.20	0.88
	(77)	(23)	(100)
1986-87	0.76	0.25	1.01
	(75)	(25)	(100)
1987-88	0.76	0.30	1.06
	(72)	(28)	(100)
1988-89	0.95	0.32	1.27
	(75)	(25)	(100)
1989-90	0.84	0.32	1.16
	(72)	(28)	(100)
1990-91	0.78	0.30	1.08
	(72)	(28)	(100)
1991-92	1.03	0.39	1.42
	(72)	(28)	(100)

Source : Directorate of Economics and Statistics, Bangalore.

Note : Figures in parentheses indicate percentages

Now sericulture has taken deep roots in almost all the taluks of the district. .Among of the eleven taluks, Kolar, Sidlaghatta, Chintamani, Mulbagal and Gowribidanur are traditional taluks of mulberry cultivation. In Bagepalle, Gudibanda, Chickbalapur, Malur, Bangarapet, Srinivasapura sericulture is moderately developed.

Though sericulture in Kolar is a fairly old and traditional activity, with the implementation of new governmental programmes in recent years there has been a considerable increase in the number of sericulturists in the district. Castewise distribution of the sericulturists in the district at the end of 1991-92, is shown in Table 6.12.

Table 6.12
CASTEWISE DISTRIBUTION OF SERICULTURISTS IN KOLAR DISTRICT (1991- 92)

Caste	Number of Sericulturists	Percentage to Total
SC	6,678	11.26
ST	3,578	6.03
OC	49,054	82.71
Total	59,310	100.00

Source : Directorate of sericulture, Bangalore.

From Table 6.12 it is clear that out of 59,310 sericulturists in the district, the majority (82.71 per cent) belonged to the catagory of Other Castes, while the Scheduled Castes formed 11.26 per cent and the Scheduled Tribes formed 6.03 per cent.

With the financial support obtained through the governmental programmes, the area under mulberry and cocoon production in Kolar has increased enormously. For the purpose of analysis, the area under mulberry and cocoon production in the district from 1980-81 to 1993-94 is presented in Table 6.13.

It is evident from Table 6.13 that during the first five years of the nineteen eighties there was almost stagnation in the area under mulberry and cocoon production. Only from 1986-87 there was a sudden upsurge both in the area under mulberry cultivation and cocoon production (Graph 6.2).

The increase in the area under mulberry hectarage and cocoon production in this district are statistically calculated and the results are given in Table - 6.14.

It is clear from the Table 6.14, that the growth rates of area under mulberry hectarage and cocoon production in the Kolar district are 4.851 per cent and 5.83 per cent respectively. The Co-efficient of determination, R^2, is 0.906 and 0.925 in both the categories implying that the growth rate is significant at 1 per cent level. The 't' values are also significant indicating a positive growth rate at 1 per cent level.

Table 6.13

AREA UNDER MULBERRY AND COCOON PRODUCTION IN KOLAR DISTRICT (1980-81 - 1993-94)

Year	Area under Mulberry (in Hectares)	Cocoons produced (in tonnes)
1980-81	16,356	9,609
1981-82	16,456	9,030
1982-83	16,456	9,766
1983-84	16,457	10,573
1984-85	16,457	10,244
1985-86	16,932	10,440
1986-87	21,417	11,889
1987-88	21,715	11,975
1988-89	22,045	12,772
1989-90	22,531	14,854
1990-91	26,181	14,643
1991-92	26,764	14,767
1992-93	26,210	18,340
1993-94	26,850	20,000

Source : Directorate of Sericulture, Bangalore.

Table 6.14

GROWTH RATES OF AREA UNDER MULBERRY AND COCOON PRODUCTION, KOLAR DISTRICT

Category	Constant Coefficient	Regression	Growth Rate	R^2	t- value
Area under Mulberry	2.6652	0.047	4.851	0.906	10.745
Cocoon Production	2.0932	0.057	5.833	0.925	12.163

Source : Computed from the data drawn from Table 6.13

The kolar district produces huge quantities of raw silk. The production of raw silk in recent years has increased as a result of improved production of reeling cocoons. The details of raw silk production between 1980-81 and 1993-94 are given in Table 6.15

It is clear from the above table that the production of raw silk has steadily increased in Kolar. It rose from 730 tonnes in 1980-81 to 2,350 tonnes in 1993-94. The growth rates of raw silk production in this district are statistically analysed and presented in the following Table 6.16.

Table 6.15

PRODUCTION OF RAW SILK IN KOLAR DISTRICT
(1980-81 - 1993-94)

(in '000 tonnes)

Year	Raw silk production
1980-81	0.73
1981-82	0.84
1982-83	0.85
1983-84	1.06
1984-85	1.03
1985-86	1.05
1986-87	1.25
1987-88	1.26
1988-89	1.34
1989-90	1.56
1990-91	1.54
1991-92	1.55
1992-93	2.04
1993-94	2.35

Source : Directorate of sericulture, Bangalore.

Table 6.16

GROWTH RATE OF RAW SILK PRODUCTION IN KOLAR DISTRICT

Constant	Regression Coefficient	Growth Rate	R^2	t- value
-0.3676	0.079	8.185	0.956	16.097

Source : Computed from the data drawn from table 6.15

It is clear from the above table that the growth rate of raw silk production is 8.185 per cent per year since 1980-81. The R^2, 0.956 and 't' value, 16.097 are also significant at 1 per cent level indicating that the growth rate in raw silk production is significant.

With the increase in the area under mulberry cultivation there was also increased demand for disease-free layings. Table 6.17 presents details about the production of disease-free layings in the government grainages of Kolar.

It is observed from Table 6.17 that there are frequent fluctuations in the production of disease-free layings. The decreasing trend in production in government srainages from 1990-91 may be due to the emergence of private grainages.

Table 6.17

PRODUCTION OF DISEASE-FREE LAYINGS IN GOVERNMENT GRAINAGES IN KOLAR DISTRICT

Year	Dfls (in lakh Nos.)
1985-86	204.64
1986-87	191.51
1987-88	206.03
1988-89	218.22
1989-90	219.67
1990-91	161.36
1991-92	154.28

Source : Directorate of Sericulture, Bangalore.

The Kolar district forms a good market-place for cocoon transactions. The details of cocoons transacted, value realised and market fee collected are given in the following Table 6.18.

Table 6.18

TRANSACTIONS MADE IN GOVERNMENT COCOON MARKETS IN KOLAR DISTRICT (1987-88 - 1991-92)

Year	Quantity of Cocoons transacted (in tonnes)	Total value realised (Rs.in lakhs)	Market fee collected (Rs. in lakhs)
1987-88	9,320	5,644	113
1988-89	7,109	5,041	100
1989-90	10,863	8,774	175
1990-91	14,223	10,198	198
1991-92	10,014	12,271	245

Source : Directorate of Sericulture, Bangalore.

The government has provided excellent infrastructural facilities for the development of sericulture in the district. The details of infrastructural facilities available in the district are given in Table 6.19.

Table 6.19 shows that there were 8 seed farms, 15 grainages, 9 markets and 169 voluntary Chawkie rearing centres in Kolar. These infrastructural facilities had their impact on the development of sericulture activity. With funds flowing from the DPAP, the NSP and other developmental programmes the district could present a sound picture of sericultural activity even in 1991.

Table 6.19

INFRASTRUCTURAL FACILITIES AVAILABLE IN KOLAR DISTRICT (1991)

SL.No.	Facilities available	No.
1.	Seed farms	8
2.	Grainages	15
3.	Chawkie rearing centres	79
4.	Voluntary Chawkie rearing Centres	169
5.	Reeling units	3
6.	Govt. Cocoon markets	9
7.	Technical service Centres	32
8.	Silk exchanges	3

Source : Office of the Dy.Director, Sericulture, Kolar.

Note:-

1. Chief Planning office, "Hand Book of statistics, 1991-92 Anantapur District", District Statistical office, Anantapur p-1
2. Ibid., p. 2
3. Office of the Deputy Director, Department of Sericulture, Anantapur.
4. Ibid., p.8.
5. Departmental note, "Growth of Sericulture in Anantapur District", Deputy Director of Sericulture, Anantapur, 1987, p-1.
6. Karnataka State Gazetteer - Kolar District, 1968, - Director of Printing, Stationary and Publications, Bangalore, p.188.
7. "Kolar District Statistical Note -1990-91", District Statistical Office, Kolar, p. 8.
8. Karnataka State Gazetteer - Kolar District Op. cit., pp. 25-26.
9. "Kolar District Statistical Note 1990-91" op. cit., pp. 13-23.
10. Karnataka state Gazetteer - Part - I, Government of India, Bangalore, 1982, p. 899.

SECTION -II

SOCIO-ECONOMIC STATUS OF SAMPLE HOUSEHOLDS

In the foregoing pages of this chapter geographical profiles of the districts of Anantapur and Kolar and the status of sericulture in them were presented. Now, in the following pages an empirical account of the sample households i.e., both sericulturists and non-sericulturists, is presented. The first part of this section deals with the socio-economic status of all the sample households and the second part covers the status of all sericulturists and non-sericulturists. It is well-known that the socio-economic status of population plays a vital role in the determination of its development. Hence, an analysis of the socio-economic status of all the sample farmers is presented here. In view of the small landholdings of majority of sample households an acre is taken as a unit instead of a hectre for the purpose of analysis.

CATEGORISATION OF FARMERS

All the sample farmers are divided into three categories depending on the size of their landholdings. The Table 6.20 gives a clear picture of this categorisation of farmers.

Table 6.20

DISTRIBUTION OF SAMPLE FARMERS AND THEIR LAND PARTICULARS IN ANANTAPUR AND KOLAR DISTRICTS (CATEGORYWISE)

	ANANTAPUR			KOLAR			
Category of the farmer	No. of Households	Total land Owned (Ac)	Average land Holding per Household (Ac)	No. of Households	Total land Owned (Ac)	Average land Holding Household (Ac)	Total
Marginal	19 (21)	31.75 (5)	1.67	250	42.06 (8)	1.68	44 (24)
Small	29 (32)	199.50 (21)	4.12	36 (40)	143.50 (27)	3.99 3.99	65 (36)
Big	42 (47)	432.20 (74)	10.29	29 (32)	348.50 (65)	12.01	71 (40)

Source : Field data

Note : Figures in parentheses show percentages.

The sample farmers are divided into three categories namely, marginal, small and big. As may be seen from the above Table, out of the 90 sample farmers in the Anantapur District, 19 are marginal owning together 31.75 acres of land, 29 small farmers with 199.50 acres, and 42 big farmers with 432.20 acres. The average landholdings of these sample farmers also vary according to the land they own and the number of farmers in that category. The average landholdings in this district works out to be 6.48 acres. The average landholdings of the marginal, small and big farmers are 1.67, 4.12 and 10.29 acres respectively.

Among the sample farmers in the Kolar district, 25 are marginal, 36 small and 29 big farmers. Out of the 90 sample farmers, 40 per cent are small farmers and form the highest percentage. Marginal farmers own 42.06 acres of land together. The small and big farmers own 143.50 and 348.50 acres respectively. The average size of the holdings in the district is 5.93 acres. The average sizes of landholdings of the marginal, small and big farmers are 1.68, 3.99 and 12.01 acres respectively.

It may be observed from the Table 6.20 that the average size of landholdings in the case of marginal farmers is almost the same both in Anantapur (1.67 Ac) and Kolar (1.68 Ac). There is a small variation in the small farmers average landholdings and almost a difference of two per cent in the case of big farmers. However, the average size of landholding (6.48 Ac) in Anantapur is higher than the average size of landholding (5.93 Ac) in Kolar.

CASTEWISE DISTRIBUTION OF SAMPLE FARMERS

Table 6.21 shows the castewise distribution of the sample farmers and their per capita landholdings.

Table 6.21 reveals that among the total 180 sample farmers, 36 belong to the Scheduled Castes and Scheduled Tribes, 30 belong to the Backward Castes and 114 belong to the Other Castes. In the Anantapur district, out of the 90 sample farmers 16 belong to the Scheduled Castes and Scheduled Tribes, 17 belong to the Backward Castes and 57 farmers belong to the Other Castes. The percentage of Other Caste farmers is higher than that of either the BCs or the SCs and STs, which is 63 per cent in Anantapur. The percentage of land owned by them is 81 per cent. The Scheduled Castes and Scheduled Tribe farmers together own only seven per cent of the land and the Backward Caste farmers own 12 per cent. The average size of landholding of the OC farmers is 8.26 acres, whereas it is 4.03 acres and 2.75 acres in the case of the Backward Castes and the Scheduled Castes and

Scheduled Tribes, respectively.

Table 6.21

CASTEWISE DISTRIBUTION OF SAMPLE FARMERS

	ANANTAPUR			KOLAR			
Category of the farmer	No. of Households	Total land Owned (Ac)	Average land Holding per Household (Ac)	No. of Households	Total land Owned (Ac)	Average land Holding Household (Ac)	Total
S & ST	16 (18)	44.00 (7)	2.75	20 (22)	82.75 (15)	4.14	36 (20)
BC	17 (19)	68.50 (12)	4.03	13 (14)	47.75 (9)	3.67	30 (17)
OC	57 (63)	470.95 (81)	8.26	57 (63)	403.56 (76)	7.08	114 (63)
Total	90 (100)	583.45 (100)	6.48	9 (100)	534.06 (100)	5.93	180 (100)

Source : Field data

Note : Figures in parentheses show percentages.

In the Kolar district also majority of the farmers belong to the Other Castes and form 63 per cent. Scheduled Caste and Scheduled Tribe farmers form 22 per cent and the Backward Caste farmers form 14 per cent. The OC farmers own 403.56 acres i.e. 76 per cent of the total land and the Backward Castes and the Scheduled Caste and Scheduled Tribe farmers own only 9 per cent and 15 per cent respectively. The average size of landholdings of the Other Castes farmers is 7.08 acres and that of the Scheduled Castes and Scheduled Tribes farmers is 4.14 acres.

The average size of landholding (4.14 Ac) in the case of Schedule Castes and Scheduled Tribes is high in Kolar district when compared to that in the Anantapur district (2.75 Ac). But the average size of landholdings of the OC farmers is high in Anantapur i.e. 8.26 acres, whereas it is 7.08 acres in Kolar.

CATEGORY AND CASTEWISE DISTRIBUTION OF SAMPLE FARMERS

The percentage distribution of sample farmers according to their category and caste is shown in the Table 6.22.

Table 6.22 shows that in Anantapur district, of the 16 Scheduled Caste and Scheduled Tribe farmers eight are marginal farmers and remaining eight are small farmers. Out of the 17 farmers who belong to the Backward Castes, five are marginal farmers and eight are small and the remaining four are big farmers. In the case of 57 Other Caste farmers six farmers are marginal 13 are small and the rest 38 are big farmers.

Table 6.22

DISTRIBUTION OF SAMPLE FARMERS (CATEGORY AND CASTEWISE)

	ANANTAPUR				KOLAR			
Category of the farmer	SC &ST	BC	OC	TOTAL	SC &ST	BC	OC	TOTAL
Marginal	8 (50)	5 (29)	6 (10)	19 (21)	6 (30)	5 (39)	14 (25)	25 (28)
Small	8 (50)	8 (47)	13 (23)	29 (32)	11 (55)	6 (46)	19 (33)	36 (40)
Big	-	4 (24)	38 (67)	42 (47)	3 (15)	2 (15)	24 (42)	29 (32)
Total	16 (100)	17 (100)	57 (100)	90 (100)	20 (100)	13 (100)	57 (100)	90 (100)

Source : Field data.

Note : Figures in parentheses show percentages.

In the Kolar district out of the 20 Scheduled Caste and Scheduled Tribe farmers six are marginal farmers, eleven are small farmers and the remaining are big farmers. Out of the 13 Backward Caste farmers, five belong to the marginal category, six belong to the small and two to the big farmers category. Out of the 57 sample farmers belonging to Other Castes, 14 are marginal, 19 are small and 24 are big farmers.

In both the Anantapur and Kolar districts, the percentage of the Other Caste farmers is more than those of the Scheduled Castes and Scheduled Tribes and Backward Castes.

LITERACY LEVELS OF RESPONDENTS

Literacy is one of the important variables which determines the status of the farmers. The following Table 6.23 gives a picture of the categorywise literacy levels of the sample farmers in both the districts.

Table 6.23.

DISTRIBUTION OF LITERACY LEVELS OF SAMPLE HOUSE HOLDS

Literacy Level	CATEGORY OF THE FARMER ANATAPUR Marginal	Small	Big	Total	KOLAR Marginal	Small	Big	Total	TOTAL
Illiterate	12 (63)	15 (52)	11 (26)	38 (42)	15 (60)	24 (66)	11 (38)	50 (55)	88 (49)
Middle School	3 (16)	7 (24)	14 (33)	24 (27)	6 (24)	5 (14)	4 (14)	15 (14)	39 (22)
High School	3 (16)	6 (21)	11 (26)	20 (22)	2 (08)	6 (17)	10 (34)	18 (20)	38 (21)
College	1 (05)	10 (03)	6 (15)	8 (09)	2 (08)	1 (03)	4 (14)	7 (08)	15 (08)
Total	19 (100)	29 (100)	42 (100)	90 (00)	25 (100)	36 (100)	29 (100)	90 (100)	180 (100)

Source: Field data.
Note : Figures in parentheses show percentage.

As seen from the Table 6.23, out of the 90 sample households in the Anantapur district, 38 farmers are illiterates. Of them 12 belong to the marginal category, 15 belong to the small farmers category, and 11 farmers are big farmers. Again 24 farmers have middle school education (three of them are marginal, seven are small and 14 are big farmers). Out of the 20 farmers who have high school education, three farmers belong to the marginal farmer category, six belong to the small farmer category and the remaining 11 belong to big farmer category. Only 8 farmers have collegiate education. (one of them is a marginal farmer, one is a small farmer and six are big farmers).

In the Kolar district out of the 90 sample farmers 50 are illiterates. Among them 15, 11 and 24 are marginal, small and big farmers respectively. Only 15 farmers have upper primary school education (17 per cent of the total literates). Eighteen farmers have high school education among whom two are marginal farmers, six are small farmers and 10 are big farmers. Only seven farmers have collegiate education and among them two are marginal farmers, one is a small farmer and the remaining four are big farmers.

In both the Anantapur and Kolar districts together, out of the total sample households only 15 farmers have collegiate education and 88 are illiterates. Thus, the illiterate farmers constitute nearly 49 per cent of the total sample. The percentage of illiterate farmers is more in Kolar than in Anantapur.

POPULATION, LAND OWNERSHIP AND WORKERS

In the Table 6.24 an analysis of the average land owned by each category of farmers, their per capita land and percentage of workers in each category is presented.

Table 6.24 reveals that in Anantapur the average family size is 5, of whom 3 (60 per cent) are workers in agricultural activities. The family size and the percentage of workers is the same in all the three categories of farmers. In this district the average land owned per household is 6.48 acres whereas the land per capita works out to be 1.29 acres.

In Kolar the average family size is 7 of whom 3 (43 per cent) are workers. Thus the percentage of workers in the family is less than that of Anantapur. The average landholding per household is 5.93 acres which is also less than that of Anantapur (6.48 AC.). Similarly the land per capita in Kolar is 0.85 acres which is far less than that of Anantapur (1.29) AC.

IRRIGATED LAND AMONG SAMPLE FARMERS

Irrigation facilities form one of the major contributors to the development of sericulture in the study area. The major sources of irrigation in the study districts are tanks, dug wells and bore-wells. Owing to erratic rain fall, the tanks remain dry for the greater part of the year. Hence, the farmers have to depend mainly on wells for their irrigation needs. The command area under each well varies from place to place and region to region. The average land under irrigation per household is shown in Table 6.25.

Table 6.25 explains that in the Anantapur district out of the 6.48 acres of average land holdings per household, an average area of 1.51 acres (23 per cent) is under irrigation. Out of the 1.67 acres of land holdings per household in the category of marginal farmers, only 0.51 acres (31 per cent) of land has irrigation facility. In the small farmers category, out of the 4.12 acres of land per household, only one acre (24 per cent) per household has irrigation facility. Out of the 10.29 acres of land per household among the big farmers 2.32 acres (23 per cent) of land per household has irrigation

Table 6.24
CATEGORYWISE DISTRIBUTION OF POPULATION, LANDOWNERSHIP AND WORKERS IN THE FAMILY

	ANANTAPUR					KOLAR				
Category of the farmer	Avarage Population per house hold (HH)	Avarage land Owned per HH (Ac)	Land per-capita in Ac	Av. number of workders per HH	Per cen-tage of Workers	Av. population on per Household HH (Ac)	Average land owned per per HH	Land per-capita in Ac	Av. number of workers per HH	Percentag of workers
Marginal	5	1.67	0.33	3	60	6	1.68	0.28	2	33
Small	5	4.12	0.82	3	60	6	3.99	0.67	3	50
Big	5	10.29	2.06	3	60	8	12.01	1.50	3	38
Pooled Average	5	6.48	1.29	3	60	7	5.93	0.85	3	43

Source : Field data.

Note : Figures in the Parentheses show the percentages.

facility. Thus the percentage of area under irrigation decreases with the increase in landholding.

Table 6.25

CATEGORYWISE DISTRIBUTION OF AREA UNDER IRRIGATION PER HOUSEHOLD

(In Acres)

Category of the Farmer	ANATAPUR			KOLAR		
	Average per Landholding household	Average Land Under irrigation per household	Percentage of Irrigated Area	Average landholding per household	Average Land Under Irrigation per household	Percentage of Irrigated Area
Marginal	1.67	0.51	31	1.68	0.78	46
Small	4.12	1.00	24	3.99	1.53	38
Big	10.29	2.32	23	13.01	3.29	27
Pooled	6.48	1.51	23	5.93	1.89	32

Source: Field Data

In the case of Kolar out of the 5.93 acres of average landholdings per household, an area of 1.89 acres (32 per cent) under irrigation. Among the marginal farmers, out of the 1.68 acres of land per household, only 0.78 acres (46 per cent) of land is under irrigation. Among the small farmers, out of the 3.99 acres of land per household, the area of 1.53 acres (38 per cent) is under irrigation. In the case of big farmers only 27 per cent of land is under irrigation. The percentage of irrigated land per household in general and in all the categories of farmers is more in the Kolar district than in the Anantapur district.

AREA UNDER PRINCIPAL CROPS

The cropping pattern followed in an area reflects its agricultural background. In Anantapur Paddy and groundnut are the predominant crops. In Kolar ragi is the chief food crop. A brief account of the croping pattern is presented in the Table 6.26.

Table 6.26 reveals that in the Anantapur district paddy and groundnut form 43 per cent of the total gross cropped area of the sample households. Ragi occupies only six percent of the area whereas the area under other crops like small millets, vegetables crops, sunflower etc., is 51 per cent of the total area. In the Kolar district, ragi and other crops occupy 75 per cent of the total cropped area. Paddy and groundnut together constitute 25 per cent of the area.

Table 6.26

CATEGORYWISE DISTRIBUTION OF PRINCIPAL CROPS

(In Acres)

Category of the farmer	ANANTAPUR					KOLAR				
	Paddy	Ground nut	Ragi	Others	Total	Paddy	Ground nut	Ragi	Others	Total
Marginal	4.80	18.95	0.50	1.15	25.40	3.25	10.35	13.55	2.60	29.75
Small	9.05	52.25	4.50	39.50	105.30	14.00	15.00	54.00	20.20	103.70
Big	40.60	66.25	20.00	190.80	317.70	41.25	20.00	75.10	145.85	282.20
Pooled	54.45 (12)	137.45 (31)	25.00 (6)	231.50 (51)	448.40 (100)	158.50 (14)	45.35 (11)	143.15 (34)	168.65 (41)	415.65 (100)

Source : Field data.

Note : Figures in parentheses show percentages.

It is interesting to note that among all the categories of farmers, groundnut takes the first place in Anantapur, whereas in Kolar ragi dominates all the other crops and in all the categories of farmers.

CATEGORYWISE PRODUCTIVITY LEVELS OF PRINCIPAL CROPS

The productivity levels of the principal crops in the study area are analysed and presented in Table 6.27

Table 6.27

CATEGORYWISE PRODUCTIVITY LEVELS OF PRINCIPAL CROPS

Per Acre Productivity Levels of Principal Crops (in Quintals)

Catogory of the Farmer	ANANTAPUR			KOLAR		
	Paddy	Ground nut	Ragi	Paddy	Ground nut	Ragi
Marginal	17.53	7.61	7.48	18.46	19.64	12.58
Small	15.47	7.89	8.86	16.63	6.87	9.03
Big	16.35	17.41	19.74	16.82	8.44	8.74
Pooled	15.86	7.62	8.69	16.86	8.19	9.19

Source : Field data.

It is evident from the table 6.27 that the productivity levels of paddy per acre in Anantapur ranges from 15.47 to 17.53, quintals. The productivity level of the marginal farmers is higher than that of big or small farmers. In the case of groundnut, the per acre productivity in all the categories of farmers is more or less the same. The average productivity is of the order of 7.62 quintals per acre. In the case of ragi there are some variations in the productivity levels among different categories.

In Kolar the productivity levels of marginal farmers are high in all the three crops. The average per acre productivity levels of paddy, groundnut and ragi are 16.86, 8.19 and 9.19 quintals respectively. However, the average productivity levels per acre of all the crops in Kolar are higher when compared with those in Anantapur.

INCOME FROM AGRICULTURE AND MILCH CATTLE

The annual income from agriculture and milch cattle in the sample districts are collected calculated and presented in the Table 6.28.

Table 6.28

CATEGORYWISE DISTRIBUTION OF PER HOUSEHOLD INCOME FROM AGRICULTURE (OTHER THAN MULBERRY) AND MILCH CATTLE PER ANNUM

Per Household Income from Agriculture and Milch Cattle

(Rs. in Thousands)

Category of the Farmer	ANANTAPUR			KOLAR		
	Agriculture	Milch Cattle	Total	Agriculture	Milch Cattle	Total
Marginal	5.31	1.10	6.41	4.30	3.50	7.80
Small	12.52	3.74	16.26	9.65	4.31	13.96
Big	26.99	5.92	32.91	30.43	12.46	42.89
Pooled	17.75	4.21	21.96	14.86	6.71	21.57
	(81)	(19)	(100)	(69)	(3.1)	(100)

Source : Field Data

Note : Figures in parentheses shows percentage.

It is clear from Table 6.28 that the income to the sample farmers from agriculture in the Anantapur district is around 81 per cent of the total income. The income from milch cattle is only 19 per cent (Rs.4.21 thousands) of the total income. The average income per household works out to be Rs.21.96 thousands. In this district the income both from agriculture and milch cattle increases along with the increase in the size of the landholdings.

In Kolar the agricultural income of the farmers forms 69 per cent of the total income of Rs.21.57 thousands. Income from milch cattle covers 31 per cent of the total income. However, the income levels increase with the increase in the size of landholdings.

The total income per household is more or less the same in both the districts. However, the income from agriculture in Anantapur district is higher than that in Kolar, whereas the income from milch cattle is higher in Kolar than in Anantapur. Moreover the income from milch cattle in all the categories of farmers was higher in Kolar, than in Anantapur. It is mainly because the Kolar district has a flurishing dairy enterprise.

COMPARATIVE ACCOUNT OF THE ECONOMIC STATUS OF SERICULTURISTS AND NON-SERICULTURISTS

In the above paragraphs the socio-economic status of all the sample respondents was presented. Their other characteristics which may influence their preference for sericulture are discussed in detail in the following paragraphs.

AREA UNDER IRRIGATION

The details of the size of holdings and the area under irrigation per household are presented in the Table 6.29.

Table 6.29

DETAILS OF THE SIZE OF HOLDINGS AND THE AREA UNDER IRRIGATION AMONG SERICULTURISTS AND NON-SERICULTURISTS

Average size holding and average area under irrigation per house hold in Ac.

	ANATAPUR		KOLAR		POOLED	
Nature of Farmers	Size of holdings	Area under irrigation	Size of holdings	Area under Irrigation	Size of holdings	Area under irrigation
Sericulturists	8.03	2.10 (26)	6.38	2.59 (41)	7.21	2.34 (32)
Non-Sericulturists	3.39	0.35 (10)	5.04	0.48 (06)	4.22	0.42 (10)

Source : Field data.

Note : Figures in paresntheses indicate percentage to size of holdings.

It is clear from the Table 6.29 that the average size of landholdings among the sericulturists is far higher than that of the non-sericulturists in both the districts. The area under irrigation of the sericulture household in Kolar is 41 per cent which is higher when compared to its counter part in Anantapur (26 per cent), whereas the percentage of area under irrigation among the non-sericulturists is higher in Anantapur (10 per cent) than it is in Kolar (6 per cent). However, on an average, the percentage of area under irrigation among the sericulturists is far more (32 per cent) than that of the non-sericulturists (10 per cent). Because of their better economic status, the sericulturists are able to invest on digging wells to provide irrigation facilities for their lands. The unit cost of one well (Dug well costs around Rs.35,000/- and that of a Bore well around Rs.60,000/-) is very high and the financial position of most of the non-sericulturists has not allowed them to invest money for this purpose.

CROPPING PATTERN

The cropping pattern of sericulturists and non-sericulturists in the sample districts is presented in the Table 6.30

Table 6.30 reveals that among sericulturists mulberry receives top priority. In Anantapur of the toral cropped area, mulberry occupies 17 per cent which is the highest among single crops. Whereas the non-sericulturists prefer groundnut.

Table 6.30

CROPPING PATTERN OF SERICULTURISTS AND NON-SERICULTURISTS

	ANATAPUR		KOLAR	
Crops	Sericulturists	Non-Sericulturists	Sericulturists	Non-Sericulturists
Paddy	44.35 (11)	10.10 (10)	39.75 (11)	18.75 (13)
Groundnut	55.80 (13)	81.65 (82)	25.50 (07)	19.85 (14)
Ragi	22.50 (05)	1.50 (02)	71.55 (20)	79.60 (51)
Others	226.00 (54)	6.50 (07)	139.35 (38)	29.30 (21)
Mulbery	70.75 (17)	-	88.50 (24)	-
Total	419.40 (100)	99.75 (100)	364.65 (100)	139.50 (100)

Source : Field data.
Note : Figures in parentheses indicate percentages.

In Kolar also mulberry takes a prominent place among single crops with 24 per cent of the total cropped area, whereas ragi dominates (51 per cent) all other crops among non-sericulturists.

INCOME FROM AGRICULTURE AND MILCH CATTLE

The incomes obtained from agriculture (other than mulberry) and milch cattle are given in the Table 6.31

The Table 6.31 shows the distributilon of gross average annual income per household from agriculture (other than mulberry) as well as from milch cattle. Among the sericulturists in Anantapur of the total income per household of Rs.26.63 thousands, agricultural income is 78 per cent and income from milch cattle is 22 per cent. Among the non-sericulturists, the total income per household is Rs.12.63 thousands, of which agricultural income forms 93 per cent and the income from milch cattle is 7 per cent.

Table 6.31

ANNUAL INCOME LEVELS PER HOUSEHOLD FROM AGRICULTURE AND MILCH CATTLE (OTHER THAN MULBERRY)

Annual income per Household (Rs. in Thousands)

Sources of Income	ANATAPUR Sericulturists	ANATAPUR Non-Sericulturists	KOLAR Sericulturists	KOLAR Non-Sericulturists
Agriculture (other than Sericulture)	20.77 (78)	11.71 (93)	17.45 (66)	14.02 (88)
Milch Cattle	5.86 (22)	0.92 (07)	9.07 (34)	2.00 (12)
Total	26.63 (100)	12.63 (100)	26.52 (100)	16.02 (100)

Source: Field data.

Note : Figures in parentheses indicate percentages.

In the case of Kolar, the average income per household among the sericulturists is Rs.26.52 thousands, of which agricultural income is 66 per cent and the income from milch cattle 34 per cent. Among the non-sericulturists, out of the total income of Rs.16.02 thousands, 88 per cent is from agriculture and the remaining 12 per cent is from milch cattle.

From the Table 6.31, it is also observed that the per household income of the sericulturists in both the districts is more or less the same, where as it is almost double the income of the non-sericulturists. It also seen from the table that the income from milch cattle is more among the sericulturists than among the non-sericulturists. This may be due to their better financial position and also because of the availability of the mulberry leaves and shoots left by silkworms which form good fodder to the cattle to increase their milk yield.

SECTION - III

ECONOMICS AND PROSPECTS OF SERICULTURE

Sericulture is a highly profitable activity and its profitability depends on the production of quality mulberry leaf and its conversion into quality cocoons at economic costs. An efforts is made here to understand the economics and prospects of sericulture, as an enterprise undertaken by the households in the Anantapur and Kolar districts. A number of variables which focus on the structure of costs and returns are presented here.

SERICULTURE ADOPTION

The sericulture enterprise is undertaken by the sample households mainly becuse mulberry is a drought resistant plant and it brings more income in shorter intervals. This is an advantage it has over other crops. Therefore the sericulture others since more than ten years. The number of farmers in each sericulture adoption group is given in Table 6.32.

Table 6.32

DURATION OF SERICULTURE ADOPTION

Duration of Sericulture Adoption	Number of Sample Sericulturists Anantapur	Kolar
Since more than 10 years	16 (27)	18 (30)
Betwen 5 to 10 years	20 (33)	16 (26)
Between 2 and 5 years	14 (23)	13 (22)
Less than 2 years	10 (17)	13 (22)
Total	60 (100)	60 (110)

Source: Field data.

Note: Figures in parentheses indicate percentages.

Table 6.32 reveals that the majority of the sample farmers in the Kolar district have been adopted this enterprise since more than ten years whereas in the Anantapur district the majority of the farmers have taken to sericulture since the last five to 10 years.

SIZE OF MULBERRY LANDHOLDINGS

The size of mulberry holdings in the study area varies from 0.25 to 6.00 acres in general. It varies from 0.25 to 2.75 acres in Anantapur and from 0.25 to 6.00 acres in Kolar. But the majority of the holdings are below 2.00 acress. Hence, the mulberry holdings are grouped into three categories viz., up to one acre, 1.00 to 2.00 acres, and more than 2.00 acres for analysing the variables that influence sericulture activity. Categorisation of the mulberry land owned by the sample farmers and the mean size of mulberry land per household is shown in Table 6.33.

Table 6.33
CATEGORISATION OF MULBERRY LANDHOLDINGS OF SAMPLE SERICULTURISTS

Size of Mulberry Holding in Ac.	Area Under Mulberry in Ac. ANATAPUR			KOLAR		
	No. of Farmers	Total Area	Average Area per Household	No. of Farmers	Total Area	Average Area per Household
Up to 1.00	39 (65)	30.50 (43)	0.78	31 (52)	24.25 (27)	0.78
1.00 - 2.00	18 (30)	32.00 (45)	1.78	22 (37)	37.25 (42)	1.69
> 2.00	3 (5)	8.25 (12)	2.75	7 (11)	27.00 (31)	3.86
Total	60 (100)	70.75 (100)	1.18	60 (100)	88.50 (100)	1.48

Source: Field data
Note: Figures in parentheses indicate percentages.

It is clear from Table 6.33 that the number of sample households decreases with the increase in the area under mulberry in both the districts. The number is more in the size group of up to one acre. It constitutes 65 and 52 per cent in the Anantapur and Kolar districts respectively. The number of cultivators in the size groups of 1.00 to 2.00 acres and more than two acres, is more in Kolar than in Anatapur. The average mulberry land per household varies from 0.78 to 3.86 acres in different groups. The pooled average mulberry land per household in Anatapur and Kolar is 1.18 and 1.48 acres respectively. Obviously the average land under mulberry cultivation per household is more in Kolar.

ESTABLISHMENT OF MULBERRY GARDEN

Mulberry being a perennial crop, the establishment of a mulberry garden is a crucial factor in sericulture enterprise and there can be no compromise on the initial establishment cost. Profitability from sericulture largely depends on the production of mulberry leaf and its conversion to cocoon production with the minimum cost. In our survey, the average size of a mulberry farm works out to be 1.18 and 1.48 acres in the Anantapur and Kolar districts respectively. Hence, one hectare as economic unit is considered as a large unit for cocoon production. Therefore, one acre is adopted as the economic unit for cocoon production in the analysis of sericulture economics. Mulberry being a leaf crop, maximisation of leaf yield depends on factors like the variety of mulberry, irrigation, application of organic manure and fertiliser, inter-cultivation etc. It is interesting to note that almost all the sample farmers in both districts have confined themselves to only two varieties of mulberry namely the Local and Kanva-2 (M_5). More than 50 per cent of the respondents use Kanva-2, nearly 25 per cent of them grow both varieties, and nearly 20 per cent use only the Local variety of mulberry. The type of plantation adopted in the study area is the row system which is popularly known as the "Kolar System". The plantation is made with cuttings. All the farms are irrigated by ground water through dug wells and bore wells. Organic manure is applied once in a year to give fertility to the mulberry farm. The Quantity of farm yard manure (FYM) varies from farmer. On an average, 17 and 16 cart loads of FYM are applied per acre in the Anatapur and Kolar districts respectively. Chemical fertiliser is applied for every crop. The data pertaining to the various operations involved in the establishment and management of one acre irrigated mulberry garden is collected and the averages obtained are presented in Table 6.34.

Table 6.34 shows the various operations and the cost involved to establish and manage one acre of irrigated mulberry garden. The establishment cost of one acre of it works out to Rs. 5165/- in Anatapur. The major items involved in the exercise are land preparation, application of farm yard manure and weeding. The mulberry garden once established is expected to give yields for 12 to 15 years. Hence, the non-recurring expenditure on the initial establishment cost is Rs. 340 and Rs. 300 per year in the Anantapur and Kolar districts respectively.

The management cost of an established mulberry garden during the first year is Rs. 4,890/- and Rs. 5,250/- in Anantapur and Kolar respectively. During the first year the number of crops harvested is small and the productivity of the crop also is low. Over the years, however, the sericulturists have gained knowledge and experience and improved their performance and productivity levels.

Table 6.34

INITIAL ESTABLISHMENT AND MANAGEMENT OF ONE ACRE IRRIGATED MULBERRY GARDEN DURING FIRST YEAR

Cost involved in Rupees

Operation	ANANTAPUR		KOLAR	
	Labour Days	Cost	Labour Days	Cost
A. ESTABLISHMENT				
1. Land Preparation	45	1125	40	1000
2. Farm Yard Manure (FYM)	17 CL	1020	16 CL	800
3. Bullock pair days involved	12	720	12	600
4. F.Y.M. Application	6	150	6	150
5. Plantation				
a. Plant Cuttings	4CL	Free	4 CL	Free
b. Cuttings preparation	4	100	4	100
c. Plantation	20	500	18	450
6. Irrigation	20	500	20	500
7. Weeding (2 times per crop)	30	750	26	650
8. Miscellaneous	--	300	--	300
Total	125	5165	114	455
B. MANAGEMENT				
1. Inter Cultivation (3 times)	60	1500	54	1350
2. Irrigation	40	1000	40	1000
3. Fertiliser Cost				
a. Complex	100kg	500	120Kg	600
b. Urea	50Kg	150	.50Kg	150
4. Fertiliser application	6	150	4	100
5. Leaf harvest	50	1250	70	1750
6. Non-recurring	--	340	--	300
Total		4890		5250

Source: Field data.

CL: Cart load.

MANAGEMENT OF ESTABLISHED IRRIGATED MULBERRY GARDEN

As already mentioned the mulberry garden once established yields for a long period. The recurring expenditure involved in the maintence of one acre established, irrigated mulberry garden is shown in Table 6.35..

Table 6.35

RECURRING EXPENDITURE ON THE ACRE OF ESTABLISHED IRRIGATED MULBERRY GARDEN FROM SECOND YEAR ONWARDS

Operation	ANANTAPUR Mandays	ANANTAPUR Cost	KOLAR Mandays	KOLAR Cost
	Cost Involved in Rupees			
1. Cultivation expenses (ploughing making ridges and furrows and weeding)	85	2125	75	1875
	12 BPD	720	10 BPD	500
2. Irrigation	40	1000	40	1000
3. Farm yard manure (FYM)	17 CL	1020	16 CL	800
4. Application of FYM	6	150	6	150
5. Cost of Chemical fertiliser				
a. 300 Kg Complex	--	1500	--	1500
b. 150 Kg Urea	--	450	--	450
6. application of Chemical fertiliser	10	250	6	150
7. Leaf harvest	125	3125	130	3250
8. Non-recuring expenditure	--	340	--	300
Total		10,680		9975

Source: Field data
CL: Cart load,
BPD: Bullock pair days.

Table 6.36

EXPENDITURE INCURRED ON MAINTENANCE OF ONE ACRE IRRIGATED MULBERRY GARDEN

(in Rupees)

Particulars	Anantapur Upto 1.00	Anantapur 1.00-2.00	Anantapur >2.00	Kolar Up to 1.00	Kolar 1.00-2.00	Kolar >2.00
	Mulberry Size Class in Ac					
1. Cultivation						
a. Human Labour	1475	1850	1875	1050	1175	950
b. Bullock Labour	500	200	200	475	190	100
2. F.Y.M.	975	825	650	880	830	620
3. Fertilisers	2150	2500	2800	2150	2100	2200
4. Miscellaneous	500	600	600	300	500	500
3. Non-recurring	340	340	340	300	300	300
Total	5940	6315	6465	5155	5095	4670
Pooled		6170			4982	

Source: Field data.

From Table 6.35, it is evident that the estimated recurring expenditure on the management of an established garden is around Rs. 10,680 and Rs. 9,975 in the Anantapur and Kolar districts respectively. With the involvement of family labour, the expenditure on mulberry cultivation can be reduced considerably. The expenditure incurred by the respondents on the maintenance of one acre irrigated mulberry garden is presented in Table 6.36.

Table 6.36 shows that the actual expenditure incurred by the sericulturists of different size groups varies from Rs. 5,940 to Rs. 6,465 in Anantapur and from Rs. 4,670/- to Rs. 5,155/- in Kolar. In fact, on an average, the farmers incur an expenditure only of Rs. 6,170 i.e., 58 per cent of the estimated cost (Rs. 10,680) in Anantapur whereas in Kolar the avaerage expenditure is Rs. 4982 which worked out to be 50 per cent of the estimated cost (Rs. 9,975).

The mulberry plantation will be fully established in one year and reaches its maximum yield from the second year onwards under irrigated conditions. The leaf yield is more or less uniform in all the seasons in the irrigated mulberry. The mulberry leaf yield obtained by the sample sericulturists is presented in Table 6.37.

Table 6.37

MULBERRY LEAF PRODUCED PER ACRE AND ITS PRODUCTION COST PER KG.

Size of Mulberry Landholding Class in Ac.	ANATAPUR		KOLAR	
	Quantity in Kg.	Cost of Production Per Kg. in Rs.	Quantity in Kg.	Cost of Production Per Kg. in Rs.
Upto 1.00	8, 006	0.74	10,700	0.48
1.01-2.00	7,164	0.88	7,522	0.68
> 2.00	10,182	0.64	8,188	0.57
Pooled	7,862	0.78	9,242	0.54

Source: Field data

It is evident from Table 6.37 that in Kolar on an average the quantity of leaf produced per acre is 9,242 Kgs which is considerably more than 7,862 Kgs produced in Anatapur. Among the mulberry cultivators of different size groups, the sericulturists of more than two acres in Anatapur and those owning upto one acre in Kolar are seen to produce more than the others.

The production cost of one kilogram mulberry leaf varies from Rs. 0.64 to Rs. 0.88 in Anatapur and from Rs. 0.48 to Rs. 0.68 in Kolar. It is interesting to note that those who produce more quantity of leaf have incurred less expenditure than those producing less.

SILKWORM REARING

Silkworm rearing is a cottage activity carried on by the sericulturist in his own house or in a separate room specially meant for it, as it requires a lot of attention. If involves a complicated process when various technical producres are to be understood and implemented for high productivity. It demands a substantial amount of mangerial skills on the part of the rearer and his own experience forms a good guide.

The rearing starts with the purchase of silkworm eggs called disease-free layings (dfls) or industrial seed normally at the cost of Rs. 225/- per hundred dfls. Both Bivoltine and Multivoltine silkworm eggs are available to the rearers. Sericulturists buy these eggs from government grainages or licensed seed-producers. In our sample all the sericulturists rear Multivoltine variety of silkworms only because of its advantage to withstand against all odds of fluctuations in temperature, unhygienic conditions in rearing sheds etc.

PLACE OF REARING

The rearing of silkworms has to take place on protected premises to avoid diseases and disturbances. The details of the rearing place of silkworms are collected from the sample households and tabulated in Table 6.38

Table 6.38

NATURE OF REARING PLACE OF SILK WORMS

Size of Mulbery Landholding in Acres	Number of Sericulturists in each category							
	ANATAPUR				KOLAR			
	1	2	3	4	1	2	3	4
Upto 1.00	4	25	10	39	5	24	6	31
1.00 - 2.00	2	12	4	18	3	12	5	22
> 2.00	2	0	1	3	9	2	2	7
Total	8	37	15	60	9	38	13	60
	(13)	(62)	(25)	(100)	(15)	(63)	(22)	(100)

Source: Field data

Note: 1. Separate rearing shed in the farm
2. Rearing in the dwelling house
3. Separate room in the dwelling house
4. Total number of farmers.

Figures in parentheses indicate percentages.

It is evident from Table 6.38 that in both districts more sericulturists (i.e., more than 60 per cent of them) rear silkworms in their dwelling houses. Only 13 per cent of the sample sericulturists in Anatapur and 15 per cent of sample sericulturists in Kolar have separate sheds in their farms for rearing. While the majority of the farmers in both districts rear silkworms in their dwelling houses. As an unfartunate consequence of this practice there is frequent occurrence of diseases infecting the silkworms.

Therefore, a separate shed is very much essential to protect the silkworms from infection and improve their quality and productivity. In the present study it is worked out that the investment on a rearing shed comes to around Rs. 25,000/- with two per cent depreciation per year, and the recurring expenditure comes to Rs. 500/- per annum towards the cost of rearing shed.

REARING EQUIPMENT

Apart from a separate shed required for silkworm rearing, it also requires a specific type of equipment which can be used for no other purpose. The details of the rearing equipment and the investment on them are presented in Table 6.39

Table 6.39

INVESTMENT ON REARING EQUIPMENT

Equipment	Total Cost in Rs.	Utility Years	Depreciation per Year in Rs.
1. Wooden rearing trays (15)	1,200	10	120
2. Chawkie rearing stand (1)	150	10	15
3. Leaf chopping board (2)	300	10	30
4. Chopping knives (2)	80	4	20
5. Antwells (36)	360	10	36
6. Rearing stands (8)	4000	10	400
7. Bamboo round trays (80)	1,600	3	533
8. Leaf chamber (1)	400	10	40
9. Bed clearing nets (160)	640	3	213
10. Foam Pads (20)	80	5	16
11. Bamboo Mountages (75)	6000	3	2000
Total	15,450		3,423

Source: Field data

It is clear from Table 6.39 that the items like wooden rearing trays, rearing stands. bamboo round trays and bamboo mountages are prominent items of fixed cost. The nonrecurring expenditure on the fixed cost of rearing equipment works out to be Rs. 3423/-. Among the sample farmers, a few rearers hire a certain number of bamboo mountages (chandrikas) on which cocoons are spun. These chandrikas are required for two or three days during the last stages of rearing. Hiring is, therefore, a matter of assembling a sufficient number of sets from the neighbours though the hiring business is not a specialised activity in these two districts.

QUANTITY OF DISEASE-FREE LAYINGS REARED

Sericulturists select a certain quantity of disease-free layings (dfls) for rearing depending on the size of their mulberry garden. The data collected from the sample households with regard to the annual quantity of dfls reared per acre, the average number of rearings undertaken in a year and the quantity reared per crop per acre are presented in the Table 6.40.

Table 6.40

NUMBER AND QUANTITY OF SILKWORMS REARED BY MULBERRY FARMERS

	Quantity Reared per Acre per Household ANATAPUR			KOLAR		
Size of mulberry Landholding Group in Ac.	**Annual Quantity (dfls)**	**Average number of rearing in a year**	**Average Quantity Reared per crop**	**Annual Quantity (dfls)**	**Average Number of Rearings in a Year**	**Average Quantity Reared per crop**
Upto 1.00	1014	3.97 (4)	255	1360	6.10 (6)	223
1.00 - 2.00	907	4.11 (4)	221	949	6.09 (6)	156
> 2.00	1290	5.0 (5)	258	1082	7.57 (8)	143
Pooled	998 (1000)	4.06 (4)	246 (250)	1120 (1100)	6.27 (6)	176 (175)

Source: Field data

Note: Figures in parentheses indicate rounded figures.

Table 6.40 shows that the average annual quantity of dfls reared per unit area of one acre is worked out to be 1,000 and 1,100 in Anatapur and Kolar respectively. However, this difference of 100 dfls between the two districts over a year is only marginal. In Anantapur, rearers who have more than two acres rear more quantity of dfls than the other two groups of rearers. In Kolar, however, the rearers of very small mulberry holdings rear more quantity of them .From the table it is clear that in both districts the rearers of one to two acres of mulberry holdigns rear less quantity than the to other groups of sericulturists.

Generally sericulturists harvest four to five crops in a year. The analysis of the rearings undertaken by the sample households reveals that on an average the rearers of mulberry holdings of more than two acres do more number of rearings in a year than the other groups. It is observed that compared with the rearings in Anantapur the number of rearings are more in all size groups of Kolar. The average number of rearings in a year are four in Anatapur and six in Kolar. Those who have been rearing more than five crops in a year divide their mulberry garden into parts to adjust with the enhanced number of rearings.

The average quantity of dfls reared per crop per acre is 250 dfls in all mulberry size groups of Anantapur, whereas in Kolar the number of dfls reared per crop decreases with the increase in mulberry area because of the pracatice of undertaking more number of rearings and thus bringing the annual total on par with others.

EXPENDITURE INVOLVED IN REARING

As analysis in the earlier paragraphs shows, the average quantity of disease-free layings reared per crop is 250 and 175 (rounded of to 200) in Anatapur and Kolar districts respectively. The estimated expenditure involved in each rearing is collected from the sample sericulturists and the averages are worked out for rearing of 200-250 dfls as presented in Table 6.41.

Table 6.41 shows that, on an average, the cost involved for rearing 200-250 dfls works out to be Rs. 2,663/- and Rs. 2,175/- in the Anatapur and Kolar districts respectively.

As already noted, the average number of rearings in a year are four in Anatapur and six in Kolar. Therefore, the expenditure on rearing per year comes to Rs. 14,572/- in Anantapur and Rs. 16,970/- in Kolar. It is inclusive of the depreciation on the rearing place and nonrecurring expenditure on the rearing equipment.

As already mentioned, the rearing activity is labour intensive and accordingly the estimated expenditure on the labour component is also higher than the other inputs involved in this activity. The involvement of family labour considerably reduces the expenditure on labour. The actual espenditure incurred by the sample rearers is indicated in Table 6.42.

Table 6.41

EXPENDITURE INVOLVED IN REARING OF 200-250 DFLS

Description	Expenditure Involved in Rupees			
	ANATAPUR		KOLAR	
	Mandays Involved	Cost	Mandays Involved	Cost
. Cost of dfls (at Rs. 225/- per 100 dfls)	250 dfls	563	200 dfls	450
2. Cost of labour wage				
a. First and Second Instars (1 labour for 9 days)	9	225	9	225
b. Third Instar (1 to 2 labour for 3 days)	6	150	3	75
c. Fourth Instar (2 to 3 labour for 4 days)	12	30	8	200
d. Fifth Instar (4 to 5 labour for 6 days)	30	750	24	600
e. Collection of ripened worms and mounting (6 to 7 labour for 2 days)	14	350	12	300
f. Cocoon harvest (3 labour for 1 day)	3	75	3	75
3. Paraffin paper formaline news paper, transportation charges etc.	--	250	--	250
Total	74	2663	59	2175

Source: Field data

Note: WAges paid at Rs. 25/- per man day.

Table 6.42

EXPENDITURE INCURRED ON REARING IN A YEAR PER ACRE

(in Rupees)

Particulars	Mulberry Size Class in acres					
	ANATAPUR			KOLAR		
	Upto 1.00	1.00 - 2.00	>2.00	Upto 1.00	1.00 - 2.00	> 2.00
Cost of dfls reared	2282	2041	2903	3060	2135	2435
Expenditure on labour	1950	1825	4125	2775	2750	1250
Non-recurring expenditure	3920	3920	3920	3920	3920	3920
Miscellaneous	900	1000	1100	1100	900	1000
Total	9,052	8,786	12,048	10,855	9,705	8,605
Pooled		9,281			9,685	

Source: Field data

From Table 6.42 it is clear that the actual expenditure incurred on rearing is Rs. 8,786/- the lowest among the rearers of one to two acres of mulberry and the highest is Rs. 12,048/- for the rearers of more than two acres in Anatapur. The average expenditure per year works out to Rs. 9,281/ -which shows a reduction of 36 per cent in the estimated expenditure on rearing in the district. In Kolar the least is Rs. 8,065/- among the rearers of more than two acres. The highest is Rs. 10,855/- among those of upto one acre. Thus, the expenditure incurred by the rearers gets reduced remarkably with the involvement of family labour.

It is also observed that in the study area those sericulturists who do not have adequate facilities for rearing young age silk worms have appraoched government and private chawkie rearing centres (CRCs). In the government chawkie rearing centres the rearing of silkworms is done free of cost whereas the sericulturists have to pay for it in private chawkie rearing centres. The users of private chawkie rearing centres have expressed different views about the services they offer and their reasons for using them. There are those users who go to the private chawkie rearing centres only for physical facilities and not for their skill or expertise. There are others, sericulturists who are very particular about the quality of dfls that can be obtained from the CRCs so that better yields could be ensured. Most of the rearers have expressed the view that they have not found any difference in the yield and cost involved between the private CRCs and their own chawkie. However, a few rearers in the Kolar district prefer buying of worms to laying from the private CRCs to avoid the risk involved in the chawkie and also to reduce the number of days of rearing activity thereby reducing the labour involvement.

COCOON YIELD

While studying the different economic parametres of sericulture we have to keep in mind the relationship between the of productivity levels and the levels of efficientcy achieved in the cultivation of mulberry and silkworm rearing. In this respect a careful uderstanding of both monetary and physical indications is essential. The levels of physical performance attained by the rearers of different mulberry size groups are analysed in terms of the yield derived by them. The cocoon yield and the performance levels of the rearers of different size groups are furnished in Table 6.43.

Table 6.43 shows that the average yield of cocoons per acre per annum is of the order of 386 and 404 Kgs, in Anatapur and Kolar respectively. The rearers of one to two acres of mulberry garden have drawn the maximum yield of 48 Kgs per 100 dfls harvested in the Anantapur district, whereas in Kolar the rearers of more than two acres of mulberry have drawn a little more yield, i.e., 49 Kgs per 100 dfls harvested. However, the average cocoon yield in both districts are more or less the same (around 47 Kgs) per 100 dfls harvested.

Table 6.43

COCOON YIELD AND PERFORMANCE LEVELS OF SERICULTURISTS IN REARING ACTIVITY

Particulars	Mulberry Size Class in acres							
	ANATAPUR				KOLAR			
	Upto 1.00	1.00 - 2.00	>2.00	Pooled	Upto 1.00	1.00 - 2.00	> 2.00	Pooled
A								
Total dfls reared per acre	1014	907	1290	1000	1360	949	1082	1100
Total cocoon yield per acre in Kgs.	375	363	516	386	454	351	431	404
Yield per 100 dfls in Kgs.	37	39	40	39	33	37	40	37
Leaf-cocoon ratio	21	20	20	20	24	21	19	23
B								
Total dfls harvested per acre	871	760	1103	848	1016	740	876	857
Percentage of harvested dfls total dfls	86	84	86	85	75	78	81	78
Yield per 100 dfls harvested in Kgs.	43	48	47	46	45	47	49	47

Source: Field data

The difference between the total dfls reared and the dfls harvested is normally accounted for by the failure of crops. The average success rates are 85 and 78 per cent in Anatapur and Kolar respectively. The failure rates of 15 per cent in Anatapur and 22 per cent in Kolar do not seem to be alarming. These failures have been caused by attack of diseases to worms during the last stages in a few cases.

The leaf cocoon ratio is a measure to identify the efficiency levels in rearing. To derive a yield of one kilogram of cocoons, the leaf used should be around 30 Kgs. Anything less than this indicates efficient use of the leaf and any thing more than this level indicates less efficient or inefficient use of the leaf. In both the districts under study the effienciency is seen to increase with the increase in the area under mulberry. On an average, the rearers of the Anatapur district show greater efficiency in the leaf cocoon ratio of 20 when compared to 23 in the Kolar district.

COSTS AND RETURNS ON ONE ACRE OF IRRIGATED MULBERRY GARDEN IN A YEAR

The success of sericulture enterprise depends on the level of profits it enjoys. The returns obtained by sericulturists act as a measure in the determining the economic viability of sericulture activity. Therefore, an attempt is made to analyse the cost and net returns from the sericulture enterprise in the study districts. Table 6.44 gives a clear picture of the costs and net returns on one acre of irrigated mulberry garden.

Table 6.44

COSTS AND RETURNS ON ONE ACRE IRRIGATED MULBERRY GARDEN

Particulars	Mulberry Size Class in Acres					
	ANATAPUR			KOLAR		
	Upto 1.00	1.00-2.00	>2.00	Upto 1.00	1.00-2.00	>2.00
A. Expenditure incurred on						
a. Production of mulberry leaf (in Rs.)	5940 (40)	6315 (42)	6465 (35)	5155 (32)	5095 (34)	4670 (35)
b. Silk worm reraring (in Rs.)	9052 (60)	8786 (58)	12,048 (65)	10,855 (68)	9705 (66)	8605 (65)
c. Total expenditure (in Rs.)	14,992 (100)	15,101 (100)	18,513 (100)	16,010 (100)	14,800 (100)	13,275 (100)
B. Quantity of cocoons produced (in Kgs)	375	363	516	454	351	431
C. Returns in Rs. (Through sale of Cocoon at Rs. 100 per Kg).	37,500	36,300	51,600	45,400	35,100	43,100
D. Net returns in Rs. (C.A.)	22,508	21,199	33,087	29,390	20,300	29,825
Pooled		Rs. 23, 150			Rs. 25,697 (or 25,700)	

Source: Field data

Note: Figures in parentheses indicate percentage to total expenditure.

It is evident from Table 6.44 that the total expenditure varies from Rs. 14,992 to Rs. 18,513 in Anantapur whereas it ranges from Rs. 13,275 to Rs. 16,010 among the different mulberry size groups. It is intersting to note the total expenditure incurred on rearing silkworms is more than the expenditure on mulberry cultivation in both the districts. It comes to nearly 61 per cent and 66 per cent of the total expenditure in the Anatapur and Kolar districts respectively. This is due to the delicate nature of the silk-worms. The rearing activity demands utmost care and skill. Hence, sericulturists have to depend to the silkworms. Therefore, the expenditure on rearing increases considerably. However, the precaustions taken result in good yield of cocoons.

With regard to net income from sericulture, the lowest is 21, 199 in the size group of 1.00-2.00 acres and the highest is Rs. 33,087 in the size group of more than two acres in Anatapur district. However, the net returns for those farmers cultivating up to one acre and 1.00 -2.00 acres of mulberry are more or less same, and the difference is very slight. In Kolar the net income per acre of irrigated mulberry is around Rs. 20,300 for the 1.00- 2.00 size class. Those who cultivate upto one acare and more than two acres get more or less the same returns. Thus, on an average, the sericulturists get a net income of Rs. 23,150 and Rs. 25,700 in the Anatapur and Kolar districts respectively.

On the basis of this per acre net returns from mulberry, an attmpt is made to analyse the income per household through sericulture. The details are furnished in the following Table 6.45.

Table 6.45

ANNUAL INCOME PER HOUSEHOLD THROUGH SERICULTURE

Mulberry Size Class (in Acres)	Annual Income Per Household in Rupees ANANTAPUR	KOLAR
Upto 1.00	17,556	22,924
1.00 - 2.00	37,734	34,307
> 2.00	90,989	1,15,125
Average	27,281	37,854

Source: Field data

It is clear from the above table that the annual income per household of differnet groups varies from Rs. 1,755 to Rs. 90,989 in Anantapur and from Rs. 22,924 to Rs. 1,15,125 in Kolar it is interesting to note that per household income is high in Kolar among those who cultivate upto 1.00

acre, though the average area per household under mulberry is 0.78 acres in both the districts. Further it may be observed that with the marginal difference in the average area under mulbery i.e., 1.18 in Anatapur and 1.48 Acre in Kolar, the average annual income from sericulture in Koalr is comparatively higher. This higher income is perhaps due to the better occupational skills of the sericulturists in Kolar.

COSTS AND RETURNS OF CROP ENTERPRISES

The respondents, in addition to mulberry, have also cultivated other crops like paddy, groundnut, ragi, vegetables, maize, bajra etc., in their lands depending on the availability of resources. Further, mulberry, paddy, groundnut and ragi are important crops grown under irrigated conditions. The comparative advantages of sericulture over other crops in terms of input costs and net returns per acre are discussed here.

INPUT COSTS OF IMPORTANT CROPS

As a result of the improvements effected in recent years, the sericulture industry in the study districts has established its economic superiority over other competing crops with a highly satisfactory cost-benefit ratio. Table 6.46 represents the data relating to the input costs of mulberry and other major crops in the sample districts.

Table 6.46 gives a clear picture of the input costs of the major crops in the sample districts. The input costs of mulberry are Rs. 15,452/- in Anantapur and Rs. 14,666/- in Kolar rather high, when compared to the input cost of other major crops like paddy, ragi and groundnut. In mulberry cultivation, the rearing of silkworms involves a major portion of the total cost. Owing to the specific nature of the rearing activity, the costs involved are always higher. But the higher level of costs is associated with higher level of returns in sericulture, whereas it is not so in the case of other major crops.The input costs of sericulture are two and half times more than that of paddy and groundnut and five times more than that of ragi.

NET RETURNS FROM SERICULTURE AND OTHER CROPS

A study of the net returns from sericulture and other major crops in the area has been undertaken for the sake of comparative analysis. Here for purposes of analysis, Paddy, groundnut and ragi are viewed as double crops and in the case mulberry the average number of crops grown per year are taken. Table-6.47 gives the details of net returns from these major crops.

Table 6.46

INPUT COSTS OF MAJOR CROPS PER ACRE

Input costs in Rupees

Operation	ANATAPUR				KOLAR			
	Mulberry	Paddy	Ground nut	Ragi	Mulberry	Paddy	Ground nut	Ragi
Cultivation Cost	1691	1950	1300	1300	1072	2300	1300	1300
Hire Charges of bullocks	330	1440	350	350	240	860	840	300
Manure and fertilizer cost	3253	1550	1250	600	2924	1400	1050	500
Seed cost	Free	240	1500	300	Free	300	1800	200
Rearing Expenditure	9281	--	--	--	9685	--	--	--
Miscellaneous Expenditure	897	500	250	250	745	500	250	300
Total	15,452	5680	5400	2800	14,666	5360	5240	2600

Source: Field Data

Note: Labour charges were paid @ 25/- per day.

Bullock pair charges were paid @ Rs. 60/- per day.

From the Table 6.47 one may gather that the pursuit of sericulture in one acre of irrigated land brings a net earning of Rs. 23148/- in Anantapur and Rs. 25,734/- in Kolar per year at the current costs of production and sale value of produce. Paddy and groundnut bring a net return of Rs.7,820/- and Rs. 10,640/- per annum in Anantapur and Kolar respectively. The net return from the cultivation of ragi in both the districts is very discouraging; it is only Rs. 700/- and Rs. 1,000/- per annum. But most of the ragi produced is used for domestic purpose. It is not for sale since it is the major food item for a majority of the farmers in these districts. The survey reveals that the net returns from sericulture in Anatapur is nearly three times that of paddy and groundnut and 33 times more than that of ragi.

In Kolar the net returns from sericulture is nearly two and half times more than that a paddy, five times that of groundnut and 26 times that of ragi. On an average, the net returns from sericulture is nearly two and half times that of paddy, three and half times that of groundnut, and twenty eight times that of ragi. The net returns from paddy, groundnut and ragi are comparatively very low. Sericulture makes an economic use of land and water, the two valuable resources and gives higher returns than paddy groundnut and ragi. It thus raises the income as well aas employment content of land thereby improving the economic conditions especially of small and marginal farmers.

Table 6.47

NET RETURNS EARNED BY THE FARMERS PER ANNUM PER ACRE FROM THE PRINCIPAL CROPS UNDER IRRIGATED CONDITIONS

Cost and Returns

	ANATAPUR				KOLAR				AVERAGE
Crop.	Yield in Physical in Rs.	Gross Value in Rs.	Expendit-ure in Rs. Terms	Net Returns in Rs.	Yield in Physical in Rs.	Gross Value in Rs.	Expendit-ure in Rs.	Net Returns in Rs.	Net Returns in Rs.
1. Mulberry cocoons	386 Kg	38,600	15,452	23,148	404 Kg	40,400	14,666	25,734	24,441
2. Paddy									
a. Paddy	38 Qtal	11,400	5,680	5,720	40 Qtal	14,000	5,360	8,640	
b. Straw	60 Qtal	2,100	--	2,100	60 Qtal	2,000	--	2,000	
c. Total	--	13,500	5,680	7,820	--	16,000	5,360	10,640	9,230
3. Ground-nut	17 Qtal	13,600	5,400	8,200	15 Qtal	10,500	5,240	5,260	6,730
4. Ragi	10 Qtal	3,500	2,800	700	12 Qtal	3,600	2,600	1,000	850

Source: Field Data

Note: Anantapur Rates:

Mulberry Cocoons = Rs. 100 per Kg.

Paddy = Rs. 300 per quintal

Ground nut = Rs. 800 per quintal

Ragi = Rs. 350 per quintal

Kolar Rates:

Mulberry Cocoons = Rs. 100 per Kg.

Paddy = Rs. 350 per quintal

Ground nut = Rs. 700 per quintal

Ragi = Rs. 300 per quintal

PRINCIPAL COMPONENT ANALAYSIS

This analaysis helps to identify the variables which can be included in a production function and it also helps to know the important elements in the activity of sericulture production. This analysis enables one to transcend the problems of conventional production function analysis, and to identify the underlying relationship between the variables.

The variables which are expected to have influence on the production of mulberry leaf and silk cocoons are presented in Tables 6.48 and 6.49 respectively. These tables present the principal components together with their associated roots. However, only three components have been presented since they together account for a bulk (69.68 per cent) of the variations. The other fifteen components together account for only 32.79 per cent.

The first component has an eigne root of 5.81 and indicates that this component accounts for 42.16 per cent of the total variation. This group is dominated by area under mulberry cultivation (0.9021), cost of manure (0.7284), cost of fertilizers (0.6938), number of layings (0.8968), lighting charges (0.7147), returns from silk cocoons (0.7753) and finally income from other crops (0.7388).

The second component yields an eigne root of 4.1630 and accounts for 18.57 per cent of variation of the variables. This component includes some of the variables which are absent in the first component. They are size of holdings (0.6962), cropping intensity (0.5820) and cost of production of other crops (0.6852).

Table 6.48

PRINCIPAL COMPONENTS - ANANTAPUR DISTRICT

VARIABLE	COEFFICIENTS		
	First Component	Second Component	Third Component in Rs.
Area under irrigation	0.4806	0.5573	-0.1942
Area under mulberry crop	0.9021	0.2485	0.0470
Landholdings size	0.5315	0.6962	-0.0859
Cropping intensity	0.2374	0.5820	0.4031
Educational levels	0.0517	0.2638	0.8106
Human labour (owned)	0.4635	-0.1293	0.0758
Human labour (Hired)	0.5842	-0.1475	0.6299
Irrigation charges	0.4560	0.2683	-0.1406
Farm yard Manure	0.7284	-0.2307	0.2518
Chemical fertilizers	06938	0.3420	-0.4693
Disinfectants	0.0764	-0.0906	0.3827
Quantity of mulberry leaves fed	0.6325	0.4712	-0.4185
Income from live-stock	0.2849	0.4381	0.1730
Number of layings	0.8968	-0.0226	-0.0649
Lighting charges	0.7147	-0.4834	-0.0870
Cost of production of other crops	0.4209	0.6852	0.3516
Returns from silk cocoons	0.7753	-0.4016	0.0625
Income from other crops	0.7388	0.2390	0.1278
Root	5.81	4.16	2.08
Percent of variation	42.16	18.57	8.95
r	0.34	0.42	0.28

Source L Field data

The third component accounts for only 7.95 per cent of the total variation in which educational level and hired labour with their values (0.8106 and 0.6299), count as the dominant variables. The remaining components are ignored since each contributes to a very negligible portion of the total variation.

Table 6.49

PRINCIPAL COMPONENTS - KOLAR DISTRICT

VARIABLE	COEFFICIENTS First Component	Second Component	Third Component in Rs.
Area under irrigation	0.4269	0.6132	0.0047
Area under mulberry crop	0.8535	0.2289	0.0605
Landholdings size	0.5071	0.7356	-0.0973
Cropping intensity	0.2640	0.5704	0.3265
Educational levels	0.1836	0.0689	0.9001
Human labour (owned)	0.3403	0.4926	-0.0852
Human labour (Hired)	0.5266	-0.2754	0.7304
Irrigation charges	0.3912	0.3586	-0.0506
Farm yard Manure	0.8640	0.2924	-0.3182
Chemical fertilizers	0.7236	0.4008	-0.1250
Disinfectants	0.0513	-0.1652	0.4299
Quantity of mulberry leaves fed	0.7602	-0.3828	0.3517
Income from live-stock	0.2426	-0.1943	0.2953
Number of layings	0.8102	-0.0161	-0.0384
Lighting charges	0.7423	-0.0850	0.0443
Cost of production of other crops	0.2690	0.7032	0.3689
Returns from silk cocoons	0.8521	-0.4156	0.1847
Income from other crops	0.7268	0.1375	0.2560
Root	7.32	4.05	1.89
Percent of variation	48.53	15.92	9.63
r	0.39	0.47	0.31

Source : Field data

Similarly for Kolar the first three components account for a large share of the total variation (74.08 per cent). The first component accounts for 48.53 per cent, the second component accounts for 15.92 per cent and the third component accounts for only 9.63 per cent, as in Anantapur.

MULBERRY LEAF PRODUCTION

To study the influence of each factor in mulberry leaf production the following Cobb-Douglas production function is employed and the model is

$$Y = \beta_0 X_1^{\beta 1} X_2^{\beta 2} \ldots X_5^{\beta 5} e^u$$

$$\text{Log } Y = \log \beta_0 + \beta 1 \log x1 + \beta 2 \log x2 + \ldots \beta 5 \log x5 + u.$$

Where :

Y= Yield on mulberry leaves in Kgs.

X_1 =Mulberry land holdings

X_2 = Human labour in man days.

X_3 = Cost of irrigation in Rupees

X_4 = Cost of manure and fertilizers

X_5 = Annual share of establishment cost of mulberry garden in rupees.

β_0 = Intercept and $\beta_1 \ldots \beta_5$ are the regression coefficients

The influence of each factor, the multiple correlation coefficient (R^2) and 'F' values are presented in the Table 6.50.

In Table 6.50 the coefficients of landholding in Anantapur, Kolar and pooled are significant at 1 per cent level indicating its + ve influence on the production of mulberry leaves and it is the only variable contributing more than the other variables. It may be noticed that the coefficients of the variable viz., human labour, cost of irrigation, cost of manure application etc., are significant at 5 per cent level in Anatapur.

Table 6.50

THE FACTOR INFLUENCE ON LEAF PRODUCTION IN ANATAPUR AND KOLAR DISTRICTS

Particulars	Anantapur	Kolar	Pooled
Mulberry Landholding(X^1)	0.6284**	0.6615*	0.6037"
	(9.53)	(5.29)	(11.71)
Human labour (X^2)	0.1073*	0.0842	0.1259**
	(2.46)	(1.67)	(2.85)
Cost of irrigation (X^3)	0.225*	0.0051	0.0076
	(2.18)	(1.14)	(1.93)
Cost of manures and fertilizers (X^4)	0.1368*	0.21 39'*	0.1842"
	(2.39)	(3.52)	(5.61)
Annual share of establishing costs of mulberry garden in Rs.(X5)	0.1950'	0.1407'	0.2364'
	(2.06)	(2.28)	(2.27)
Intercept	12.8316	9.5362	9.7025
Multiple correlation coefficient (R2)	0.6259	0.6803	0.6140
F - ratio	84.31*'	121.75'*	92.50"

Source: Field data

Note : * : Significantat 5percent level

** : Significant at 1 per cent level

In the case of Kolar the coefficient of the cost of fertilizer is significant at 1 per cent and the annual share of establishment costs of mulberry garden is significant at 5 per cent level only. The other two have no significant contribution to the mulberry leaf production. The multiple correlation coefficients are also significant and are presented in the table indicating that the fit is a good one.

The above R^2- values reveal that more than 60 per cent of the variation is explained by the five variables.

SILK COCOON PRODUCTION

The same Cobb-Douglas production function is used to study the influence of each factor on cocoon production. The model used is as follows:

$$Y = ß_0 X_1^{ß_1} X_2^{ß_2} X_3^{ß_3} X_4^{ß_4} X_5^{ß_5} e^u$$

Where Y = Yield of silk cocoons in Kgs.
X_1 = Mulberry leaves fed in Kgs.
X_2 = Number of disease-free layings.
X_3 = Human labour in mandays.
X_4 = Disinfectant charges.
X_5 = Lighting charges.

$ß_0$ is an intercept and $ß_1 \ldots ß_5$ are regression coefficients.

The regression coefficients of the above variables, R^2 and 'F' values are presented in the Table 6.51

It is evident from the Table 6.51 that the coefficients of the number of disease-free layings in Anantapur, Kolar and pooled are significant at 1 per cent level implying their + ve influence on the production of cocoons. It is the only variable which contributes more than the other variables. In the case of the other variables, mulberry leaves, disinfectant charges and lighting charges, the pooled coefficients are significant at 5 per cent level. In Kolar the cost of disinfectants is significant at 5 per cent level. The other variables show no impact on production of cocoons. The multiple correlation coefficients reveal that more than 75 per cent of variation is explained by the three variables.

Table 6.51

THE FACTOR INFLUENCE ON SILK COCOON PRODUCTION IN ANANTAPUR AND KOLAR DISTRICTS

Particulars	Anantapur	Kolar	Pooled
Mulberry leaves in Kgs (X_1)	0.0625	0.0293	0.0741
	(0.83)	(1.56)	(2.18)
Number of disease free layings (X_2)	0.1 073**	0;71 58**	0.6392*'
	(5.24)	(9.36)	(3.78)
Human labour employed mondays (X_3)	0.0847	-0.0264	0.0488
	(1.62)	(1:90)	(1.06)
Disinfectants charges in rupees (X_4)	0.03905	0.1057*	0.1 623*
	(0.98)	(2.45)	(2.11)
Lighting charges in rupees (X_5)	0.1 926	0.0738	0.1259
	(1.37)	(1.82)	(1.37)
Intercept	-4.7319	-0.5603	2.3061
(R_2)	0.8246	0.7529	0.7840
F-ratio	43.18	129.62	30.96

Source: Field Data.

Note: * Significant at 5 per cent level.

** Significant at 1 per cent level.

CONCLUSION

Sericulture has struck deep rootes in the Anantapur and Kolar districts of Andhra Pradesh and Karnataka respectively. It is practiced in almost all parts of these districts. In Kolar sericulture is regarded as a traditional occupation. The two districts have a unique distribution of similar agroclimatic conditions because of their geographic locations. Mulberry cultivation in these two districts is being undertaken under irrigated conditions. Mulberry being drought resistant, recurring droughts in these two districts have not pervented the tremendous growth of Sericulture. Thus, they have become pioneers in sericulture in their respective states. The development of Sericulture in these two districts is largely attributed to the efforts of their respective Government Departments of Sericulture.

Sericulture in Anantapur was first introduced in Talamarla a village formerly of the Kadiri taluk during 1923-24. It had to face teething problems until the commencement of the Drought Prone Areas Programme during 1975. Sericulture in the district gained momentum with the implementation of the DPAP and by the end of 1993-94, the area under mulberry went up to 33,220 hectares with the cocoon production of 11,470 tonnes. In Kolar, with the implementation of governmental programmes sericulture gained momentem and by the end of 1993-94 the area under mulberry went up to 26,850 hectares with the cocoon production of 20,000 tones. The district has adequate in frastructural facilities to meet the present demand.

A comparative account of the socio-economic status of sericulturists and non-sericulturists in both the study districts shows that the sericulturists are better placed than the non-sericulturists. The average land holding and the percentage of area under irrigation are more among sericulturists because of their better financial position. The cropping pattern shows that the non-sericulturists give priority to groundnut in Anantapur and to ragi in Kolar, whereas the sericulturists in both the districts give priority to mulberry. The annual income per household reveals that the income from agriculture (other than sericulture) as well as from milch cattle is much higher among the sericulturists.

To understand the sericulture enterprise, as carried out by different size groups of mulberry landholdings, a number of variables which focus on the structure of costs and returns have been examined. The average mulberry holdings per household in Anantapur 1.18 acres which is less when compared to that of 1.48 acres in Kolar. The initial establishment cost of one acre mulberry garden is Rs. 4550 in Kolar and Rs. 5165 in Anantapur. The recurring expenditure on one acre of an established irrigated mulberry garden is estimated to be around Rs. 10,000 per annum. This cost is reduced considerably by Rs. 5 to 6 thousands with the involvement of family labour. The cost of production of one kilogram mulberry leaf is worked out to be Rs. 0.54 in Kolar and Rs. 0.78 Anantapur.

Majority of the sample households in the study area have undertaken the rearing activity in their own dwelling houses. On an average, 1000 to 1100 disease-free layings are reared and the average quantity reared per crop per acre is 200 to 250 dfls. It is observed that on an average 4 to 6 rearings in a year are undertaken by the sample households. All the sericulturists rear multivoltine variety of silkworms only, since this type of silkworms can withstand the fluctuations in temperature and ensure high cocoon yield. The expenditure involved in a year in the rearing activity is estimated to be Rs. 14,572 and Rs. 16,970 in Anantapur and Kolar respectively. The involvement of family labour significantly reduces the costs of mulberry cultivation and silkworm rearing. On an average, the cocoon yield is around 47 kg per 100 dfls. The analysis of costs and returns shows that the net returns per acre in a year are of the order of Rs. 23,150 and Rs. 25,700 in the Anantapur and Kolar districts respectively. The average annual income per household through sericulture is Rs. 27,281 in Anantapur and

Rs. 37,854 in Kolar. Thus sericulture contributes significantly to the total income of the sample households.

Though sericulture, appears initially to be an expensive enterprise when compared to the other principal crops like paddy and groundnut (three to six times more), the net returns from mulberry are found to be much higher than the return from paddy and groundnut (four to six times). Sericulture exacts higher employment potential and enhances the income levels, and thereby improves the economic status of the farmers.

The Principle Component Analysis which is used to identify the variables, which influence the production of mulberry leaves reveals that the first three variables, area under irrigation, area under mulberry crop and the size of the landholdings, account for a larger share of the total variation in both the districts. The Cobb-Douglas production function also reveals that the pooled co-efficients of landholdings are significant at 1 per cent level indicating positive influence on the production of mulberry leaves. The same Cobb-Douglas production function which is employed to study the influence of each factor on cocoon production shows that the co-efficients of the number of disease free layings is significant at 1 per cent level and it is the only variable which contributes more to cocoon production than other variables.

CHAPTER 7

Employment Generation Through Sericulture

Sericulture, as already mentioned, generates a chain of economic activities providing employment opportunities to the people in rural, urban and semi-urban areas. Through sericulture the incidence of unemployment, disguised unemployment and seasonal unemployment is relieved to some extent, especially in the agricultural sector. Employment opportunities in sericulture activity can be classified according to their nature, into agricultural and industrial. Mulberry cultivation, silkworm rearing and cocoon production are agricultural and can be taken up in rural areas. Silk reeling, twisting, weaving, dyeing etc., are industrial activities and can be undertaken in urban and semi-urban areas. In this chapter, an attempt is made to analyse the employment potential of sericulture covering mulberry cultivation and silkworm rearing.

Human labour involvement in mulberry cultivation is categorised into two parts. One is employment generation in the first year of establishment of the mulberry garden. The other is creation of labour days from the second year onwards. The data relating to them in the study districts are presented in the following pages.

HUMAN LABOUR INVOLVEMENT DURING ESTABLISHMENT YEAR

Human labour involvement in sericulture during the first year covers the establishment of the mulberry garden, its management and silkworm rearing. The employment generated through these activities is presented in Table 7.1.

Table 7.1 reveals that the establishment of one acre mulberry garden generates 281 and 252 man days in Anantapur and Kolar respectively. In Anantapur the number of man days generated is slightly more because this enterprise is new to the farmers in this district. The management of mulberry garden during the first year creates 156 man days in Anantapur and 168 man days in Kolar. During the rearing of silkworms 222 and 177 man days are generated in Anantapur and Kolar respectively. However, in total during the establishment year 659 man days in Anantapur and 597 man days in Kolar are generated.

Table 7.1

HUMAN LABOUR INVOLVEMENT ON ONE ACRE OF MULBERRY DURING THE FIRST YEAR OF ESTABLISHMENT

Operation	Number of Man days generated		
	Anantapur	Kolar	Pooled
I MULBERRY			
A. Establishment			
1. Cultivation	105	94	100
2. F.Y.M. application	6	6	6
3. Preparation of cuttings	4	3	4
4. Plantation	20	15	18
5. Irrigation	60	55	58
6. Weeding	30	25	28
7. Chemical Fertiliser application	6	4	5
8. Leaf harvest	50	50	50
Total	281	252	268
B. Management			
1. Inter Cultivation	60	54	57
2. Irrigation	40	40	40
3. Fertiliser application	6	4	5
4. Leaf harvest	50	70	60
Total	156	168	162
II. SILKWORM REARING			
1. First and Second Instars	27	27	27
2. Third Instar	18	18	18
3. Fourth Instar	36	24	30
4. Fifth Instar	90	60	75
5. Collection of Riped Worms & Mounting	42	42	42
6. Worm harvest	9	6	8
Total	222	177	200
Grand Total	659	597	631

Source : Field data.

HUMAN LABOUR INVOLVEMENT IN AN ESTABLISHED MULBERRY GARDEN

The employment opportunities generated in an established mulberry garden are given in Table 7.2

Table 7.2

HUMAN LABOUR INVOLVEMENT ON ONE ACRE ESTABLISHED MULBERRY GARDEN

Operation	Number of mandays generated		
	Anantapur	Kolar	Pooled
I MULBERRY GARDEN			
1. Cultivation	85	75	80
2. Irrigation	40	40	40
3. Farm Yard Manure application	6	6	6
4. Chemical Fertiliser application	10	6	8
5. Leaf harvest	25	130	128
Total	266	257	262
II SILKWORM REARING			
1. First and Second instars	36	54	45
2. Third instar	24	36	30
3. Fourth instar	48	48	48
4. Fifth instar	120	120	120
5. Collection of riped worms & Mounting	56	84	70
6. Worm harvest	9	12	11
Total	296	354	324
Grand Total	562	611	586

Source: Field data.

It is clear from Table 7.2 that from an established mulberry garden 266 man days in the Anantapur district and 257 man days in the Kolar district are generated. Leaf harvest and cultivation of the mulberry garden are major items which involve considerable human labour. Cultivation includes operations like deep ploughing, making ridges and furrows, weeding etc. The generation of man days in an established mulberry garden in the two study districts is more or less the same, the difference being minimal.

In the entire activity of sericulture, the next important aspect in the generation of employment opportunities is the silkworm rearing. Rearing is a household activity and has more employment potential than mulberry cultivation. As already mentioned, on an average 250 - 300 dfls can be reared on one acre of mulberry garden. But in the study sample, the sericulturists have been able to rear 245 dfls (250 dfls), and 175 dfls (200 dfls) per acre, per crop in the Anantapur and Kolar districts respectively. The rearing of silk worms generally takes 24-28 days. The sample sericulturists in Anantapur make four rearings per year whereas the sericulturists in Kolar generally achieve six rearings per year. The number of man days generated per year in these two districts are calculated taking the average number of rearings per year. It works out to be 296 man days and 354 man days in Anantapur and Kolar respectively. However, the above analysis clearly indicates the employment potential of silk worm rearing.

Further, there is a lot of variation in the generation of working days in rearing silkworms in Anantapur and Kolar. The practice of sericulture is a new enterprise to the farmers in Anantapur district. Though they rear 245 dfls per crop, the generation of man days per year is 296 only, since the average number of rearings are only four. Sericulture, in Kolar is a deep-rooted and traditional occupation for the past 200 years. As most of the farmers in this district are well-versed with the practice of sericulture, within the available one or two acres of land, they are able to cultivate mulberry on the rotation system so that some times they are able to achieve 8 - 10 rearings per year. Hence, the average number of rearings is higher in Kolar. In the study sample 354 man days out of six rearings are generated on the rearing of 200 dfls per crop.

Inspite of slow progress of sericulture in Anantapur, the farmers in the district have taken to it enthusiastically because of its various advantages although the investment involved is rather high compared to the number of man days generated and the number of rearings per year is less. Both activities, mulberry cultivation and silk worm rearing, create 562 man days and 611 man days in Anantapur and Kolar respectively. The employment opportunities per household generated are calculated and presented in Table 7.3.

From Table 7.3, it is evident that the generation of mandays per household is more in the group of sericulturists who own more than two acres of mulberry land. It works out to 1,546 and 2,358 man days in Anantapur and Kolar respectively. Man days generated in the class of those who own upto one acre are 438 and 477 and in the class of those who own one to two acres, they are 1,000 and 1,033 in Anantapur and Kolar districts respectively. In total, among the sample sericulturists 39,720 man days in Anantapur and 54,019 man days in Kolar are generated per year. The pooled averages of these are 662 and 900 man days respectively. The generation of man days in each class is more in the Kolar district because of the reasons already mentioned.

Table 7.3

EMPLOYMENT GENERATION PER HOUSEHOLD THROUGH SERICULTURE

Mulberry Size Class in Ac	Man days generated ANANTAPUR		KOLAR	
	Total	Per House-hold	Total	Per House-hold
Upto 1.00	17,082	438	14,787	477
1.00 - 2.00	18,000	1,000	22,726	1,033
>2.00	4,638	1,546	16,506	2,358
Pooled	39,720	662	54,019	900

Source: Field data.

From Table 7.3, it is evident that the generation of mandays per household is more in the group of sericulturists who own more than two acres of mulberry land. It works out to 1,546 and 2,358 man days in Anantapur and Kolar respectively. Man days generated in the class of those who own upto one acre are 438 and 477 and in the class of those who own one to two acres, they are 1,000 and 1,033 in Anantapur and Kolar districts respectively. In total, among the sample sericulturists 39,720 man days in Anantapur and 54,019 man days in Kolar are generated per year. The pooled averages of these are 662 and 900 man days respectively. The generation of man days in each class is more in the Kolar district because of the reasons already mentioned.

MAN DAYS GENERATED IN COMPARISON WITH SOME IMPORTANT CROPS

Employment generation through other important crops in the sampe districts is analysed for the purposes of comparison. The cropping pattern in Anantapur and Kolar reveals that Paddy, Groundnut and Ragi are important commercial food crops, which occupy a prominent place in the agricultural economy of these two districts. Therefore, a study of the employment opportunities generated through them in comparison with those created by sericulture is sure to be interesting. Generally while analysing the labour units required for a crop due consideration is given to the difference in soils and agricultural practices. Necessary scientific enquiry should also be made regarding the fertility of the soil, customs and practices etc. But for determining the employment potential of different crops, such a study may not be necessary. What is required is an exercise, which would give a board idea of the quantity of employment generated by each crop. A micro level study of this aspect reveals that there is far more scope for employment generation in sericulture than in the other commercial crops. In the present analysis, paddy, groundnut and ragi are taken as double crops and mulberry as an annual crop. Sugarcane, however, is not included in the analysis as it is not cultivated in the sample villages. Table 7.4 presents the labour requirement in terms of man days for the four major crops per acre per year under irrigated conditions.

The Table 7.4 reveals the generation of man days on one acre of mulberry garden and other important crops. In Anantapur the generation of man days in one acre of mulberry garden under irrigated condition is three times more than that of paddy, nearly four times that of groundnut and more than five times that of ragi. In comparison with these figures in Anantapur, the generation of labour days in Kolar shows some variations. In this districst the number of man days generated in one acre of mulberry farm under irrigated conditions is nearly four and half times more than that of paddy and groundnut, and five times that of ragi. Though paddy, groundnut and ragi are taken as double crops the labour requirements for them are considerably less than far sericulture. The average involvement of labour in the Anantapur and Kolar districts is shown in the Table 7.5.

Table 7.5 gives a clear picture of the number of man days generated in one acre of irrigated land in the sample districts. The average number of man days generated in the mulberry farm is three and half times more than that of paddy, four times that of groundnut and five times that of ragi in the sample districts. This analysis clearly proves the superiority of sericulture over the other important crops in the generation of employment opportunities especially in rural areas where people depend on agriculture.

The above table shows that the calculated 't' values are significant at 1 per cent level indicating that there is a significant increase in employment generation due to sericulture adoption in both the districts. Further, the above results reveal that employment generation is better in Kolar than in Anantapur.

Table 7.4

LABOUR INVOLVEMENT OF ONE ACRE OF MULBERRY, PADDY, GROUNDNUT AND RAGI UNDER IRRIGATED CONDITIONS

Operation	Labour involvement in Man days ANANTAPUR				KOLAR			
	Mulberry	Paddy	G. Nut	Ragi	Mulberry	Paddy	G.Nut	Ragi
1. Nursery / Raising (Cultivation in the case of Mulberry)	85 12 BPD	4 2 BPD	20 4 BPD	2 --	75 10 BPD	4 2 BPD	24 6 BPD	2 --
2. Main field preparation	--	6 20 BPD	-- 8 BPD	2 8 BPD	-- 8 BPD	6 2 BPD	-- 6 BPD	2 8 BPD
3. Transplanting	--	30	--	16	--	30	--	18
4. Manure and fertiliser application	16	6	8	2	12	8	8	2
5. Weeding (In the case of mulberry it is included in cultivation)	--	40	40	24	--	48	32	26
6. Plant protection	--	4	4	4	--	4	4	4
7. Irrigation	40	30	30	24	40	30	24	28
8. Harvesting	125	50 2 BPD	40 2 BPD	30 2 BPD	130	40 1 BPD	36	34 2 BPD
9. Rearing of silkworms	296	--	--	--	354	--	--	--
Total	562+ 12 BPD	170+ 24 BPD	142+ 14 BPD	104+ 10 BPD	611+ 10 BPD	136+ 10 BPD	128+ 10BPD	116+ 10BPD

Source: Field data.
Note: BPD = Bullock pair days.

Table 7.5

THE AVERAGE NUMBER OF MAN DAYS GENERATED IN MULBERRY AND OTHER CROPS PER YEAR (PER ACRE)

Name of the crop	No. of man days generated		
	Anantapur	Kolar	Average
Mulberry	562	611	586
Paddy	170	136	153
Groundnut	142	128	135
Ragi	104	116	110

Source: Field data.

Table 7.6 indicates the significance of sericulture in the creation of employment opportunities in the Anantapur and Kolar districts.

Therefore, the above analysis makes it amply clear that mulberry cultivation and silkworm rearing generate more number of working days than the other crops to the rural people. Apart from this, sericulture beneficially engages a large number of women and old people in the family in rearing silkworms. The participation of women at the time of feeding the young age silkworms and at the times of transferring the ripe silkworms from trays to chandrikas is high. In this way sericulture serves as a most suitable cottage industry in providing employment opportunities to women and aged people with minimum risk. Moreover, employment opportunities can also be created indirectly in the ancillary sectors of sericulture. For example, employment opportunities are created in the manufacturing of rearing equipment required for rearing silkworms'. Further, silk worms waste can be used in manufacturing oils, cakes and sometimes as cattle feed. All these create employment opportunities indirectly. Hence, sericulture has significant role to play in generating employment opportunities and therefore is directly relevant to efficient policy formation.

Table 7.6

COMPARISION OF MEANS ON EMPLOYMENT GENERATED BETWEEN ANANTAPUR AND KOLAR DISTRICTS BY 't'-TEST

District	Standard deviation	t-value
Anantapur	2.2017	13.5279
Kolar	1.8623	21.8460

Source: Field data.

CONCLUSION

The analysis on generation of employment opportunities in sericulture reveals the superiority of mulberry cultivation and silk worm rearing over other crops in the sample districts. The establishment and maintenance of mulberry garden, which is not technical in nature, involves a large number of labour days. Silkworm rearing is a delicate operation to be done with utmost care and attention, and therefore requires more number of working days. Silk worm rearing in the Kolar district involves 59 mandays per rearing where as in the Anantapur district 74 are required per each rearing. On an average 296 man days and 354 man days per year are generated in Anantapur and Kolar respectively. As sericulture is a traditional occupation in Kolar, silkworm rearing can be taken up very easily and managed skillfully by family members. Sericulturists in Kolar are also able to undertake more number of rearings than their counterparts in Anantapur. But in Anantapur where the situation is entirely different, sericulturists mainly depend on skilled labourers for silkworm rearing. Owing to this reason, the creation of man days per each crop is more in this district.

However, sericulture is gradually gaining the status of a main crop by taking precedence over other traditional crops in creating employment opportunities. The job opportunities created by it are far more than those by the traditional crops such as paddy, groundnut and ragi. The number of man days created in one acre mulberry farm under irrigated conditions is three and half times more than that of paddy, four times that of groundnut and five times that of ragi in the sample districts.

Sericulture activity also creates indirect employment opportunities in ancillary industries which are necessary for the development of sericulture. The by-products of sericulture further generate employment opportunities. Therefore, sericulture is a suitable tool to eradicate seasonal and disguised unemployment in rural areas as it can generate both on-farm and off-farm employment.

CHAPTER 8
Problems of Sericulture in India

The study of sericulture development in the Anantapur and Kolar districts, presented in the preceding chapters, shows that sericulture has made quick and remarkable progress during the past 15 years, as small and marginal farmers in the area are being attracted more and more to this activity because of the prospects it holds. Being agrobased, it has vast potential for generating income and employment opportunities primarily to the rural masses. This potential appears to have further increased with governmental intervention to promote and achieve an all round development of sericulture with the help of Five-Year programme financed by the World Bank.

There is, no doubt, an increase in the production of cocoons relieved the silk industry from the problem of scarcity of raw material. But the farmer who grows mulberry and rears silkworms faces certain other problems in the operation of sericulture activity. Unless these problems are identified and necessary action taken, the programme of planned sericulture development may not succeed and the targetted growth can not be achieved. This chapter attempts to highlight the factors which come in the way of sericulture development in the Anantapur and Kolar districts.

PROBLEMS OF SERICULTURE

The sericulture industry has been suffering from many problems pertaining to its organisation, marketing, finance, technology and extension. Depending on the intensity of the problems encountered by the sample sericulturists, their perceptions are collected, ranked and presented in the Table 8.1.

From Table 8.1 it is evdent that all the respondents in the Anantapur district have felt the problems of pests and diseases as severe. Next in order, 92 per cent of them have felt actully the problem of non-availability of layings during summer. Nearly 87 per cent of them find fluctuations in Cocoon prices problematic. The problem of shortage of labour, climatic disturbances, shortage of skilled workers are experienced by as many as 83 per cent, 78 per cent and 75 per cent respectively. About 70 per cent of the sample farmers have felt the problem of inadequate finance for investment. Shortage of rearing equipment and inadequate market facilities come next in ranking. Nearly 18 per cent of the farmers are faced with the problem of inadequate demand for the cocoons. This problem may have arisen because the demand for cocoons for a certain period has decreased because of imports from China.

In the Kolar district too all the respondents as in Anantapur, find pests and diseases as their foremost problem, but their response to other problems has been different. Fluctuations in cocoon prices is their next important problem for nearly 87 per cent of the sample farmers there. About 83 per cent of the farmers face the problem of shortage of rearing equipment, and nearly 83 per cent have felt the problem of shortage of skilled labour. Next in order of importance appear problems of inadequate finance for investment (70 per cent) and shortage of labour (67 per cent). In the list of problems faced by them are securing extension officers' advise (50 per cent), non-availability of layings (42 per cent), and inadequate demand for cocoons (37 per cent). Fortunately not even one of them refers to the problem of inadequate markeing facilities, because they have access to the famous silk markets at Sidlaghatta, Kankapura, Chintamani, Vijayapur and Ramanagar, which are not very far from Kolar district. However, 17 per cent of them speak of the problem of transportation of cocoons.

REASONS FOR NON-ADOPTION OF SERICULTURE

There are farmers in both the study districts who have not adopted sericulture inspite of their knowing its various advantages. The reasons given by them for not adopting it are ranked and presented in Table 8.2.

From Table 8.2 it may be known that 93 per cent of the sample non-sericulturist farmers in Anantapur and all the sample non-sericulturist farmers in Kolar are hesitant to take up sericulture due to the fear of diseases infecting silkworms. All of them are under the impression that frequent occurance of diseases among the silkworms will not only affect their economic prospects, but once the disease infects the worms, the entire crop will be lost. If the extension personnel of the Department of Sericulture provide them with the necessary knowledge about the control of pests and diseases, and make available to them the disinfection measures etc., they may overcome their fear and come forward to cultivate mulberry.

Table 8.1

RESPONSE OF SAMPLE FARMERS ABOUT THE PROBLEMS OF SERICULTURE

Sl. No.	PROBLEMS	ANANTAPUR		KOLAR	
		No. of Farmers	Rank	No. of Farmers	Rank
1.	Pests and diseases	60 (100)	I	60 (100)	I
2.	Non-availability of layings	55 (92)	II	25 (42)	VIII
3.	Fluctuation in cocoon prices	52 (87)	III	52 (87)	II
4.	Shortage of labour	50 (83)	IV	40 (67)	VI
5.	Climatic Disturbances	47 (78)	V	20 (33)	X
6	Shortage of skilled workers	45 (75)	VI	45 (75)	IV
7	Inadequte finance for investment	42 (70)	VII	42 (70)	V
8	Shortage of rearing equipment	40 (67)	VIII	50 (83)	III
9.	Inadequate market facilities	36(60)	IX	0(0)	XIV
10.	Securing Extension Officers Advise	32 (53)	X	30 (50)	VII
11	Exploitation by the middle men	30 (50)	XI	32 (37)	IX
12	Transportation problem	25 (42)	XII	10 (17)	XIII
13	Inadequate demand for cocoons	11 (18)	XIII	11 (18)	XII
14	Shortage of pesticides	10 (17)	XIV	15 (25)	XI

Source: Field data.

Notes : Figures in parentheses indicate the percentages of farmers.

The other reason specified by the non-sericulturists in Anantapur is the lack of irrigation facilities. As many as 83 per cent of the farmers raise this problem. Next in order come the need for investable funds and non-availability of layings whenever they are required. Banks and co-operatives may help to remove this fear by raising the unit cost of Rs.19,800/- , so as to enable the farmers to switch over from the traditional crops to sericulture. Timely availability of layings should also be attended immediately. Eighty per cent of the sample non-sericulturist farmers admit that they have not taken to sericulture due to a lack of knowledge about desinfection measures. Next in order are such problems as lack of land for the construction of rearing house, shortage of skilled and other labour, and non-availability of mulberry cuttings. Only 33 per cent of the farmers expressed their fear of fluctuations in cocoon prices and demand for cocoons.

The non-sericulturists in the Kolar district also give the same reasons for not taking to sericulture but their ranking is different. All of them are apprehensive of diseases infecting the silkworm. This fear commonly shared by the non-sericulturists in both the study districts. The second reason given by the Kolar non-sericulturists (87 per cent) is the lack of funds for investment and their third reason is the shortage of skilled workers. Fourth and fifth reasons are lack of knowledge about disinfection and lack of irrigation facilities.

In spite of the number of problems in sericulture farming, the motivating factor for taking up sericulture is the growing demand for silk in both national and international markets. However a brief account of the major problems ecountered by sericulturists and measures to be taken to overcome them are presented in the following paragraphs.

INTENSITY OF PESTS AND DISEASES FOR SILK WORMS

Mulburry cultivation, the first step in producing silk, is an easy job and does not involve any special techniques or require special knowledge. Mulberry is affected very rarely by leaf rust or leaf spot diseases. Though there are some measures to control these pests, the only measure followed by the sample sericulturists is separating the pest affected leaves from the plant and destroying them.

The major operation of sericulturists is silkworm rearing which needs the particular care and attention of the rearers. It is the most delicate activity in the entire operation of silk production.

Table 8.2

REASONS FOR NON-ADOPTION OF SERICULTURE BY THE NON-SERICULTURISTS

Sl. No.	PROBLEMS	ANANTAPUR		KOLAR	
		No. of Farmers	Rank	No. of Farmers	Rank
1	Fear of disaeses	28 (93)	I	30 (100)	I
2	Lack of irrigation facilities	26 (86)	II	20 (67)	V
3	Lack of capital for investment	25 (83)	III	26 (87)	II
4	Non-availability of layings in righttime	25 (83)	IV	15 (50)	IX
5	Lack of knowledge about disinfection measures	24 (80)	V	22 (73)	IV
6	Lack of land for construction of rearing house	22 (73)	VI	20 (67)	VI
7	Shortage of skilled labourers	20 (67)	VII	23 (77)	III
8	Shortage of labourers	20 (67)	VIII	15 (50)	X
9	Non-availablity of mulberry cuttings	15 (50)	IX	10 (33)	XI
10	Lack of capital for construction of rearing house	15 (50)	X	16 (53)	VIII
11	Non-availability of rearing equipment (for hire)	12 (40)	XI	18 (60)	VII
12	Fluctuations in cocoon prices	10 (33)	XIV	8 (27)	XII

Source: Field data.

Note : Figures in parentheses indicate the percentages of farmers.

Grassarie, Flacherie, Pebrine and Uzyfly are the enemies of silkworms and silkworm rearers, which decrease the quantity and quality of cocoons. Therefore as a precautionary measure, silkworm rearers have to use nylon nets and have the rearing equipment properly cleaned with formaline, and also use Uzycide and 'Reshamkeet Oushad' as preventive measures. Once the silkworms are attacked by Grasserie, Flacherie or Uzyfly, the infected silkworms should be destroyed at once. There are nc control measures to prevent these diseases except conducting the rearing operations under the most hygienic conditions.

Since frequent occurrence of diseases to silkworms is one of the major problems troubling sericulture some new farmers who want to adopt sericulture are scared of these pests and diseases and hesite to take it up. So, the research organisations are seized of this problem, and are trying seriously to find a remedy for it. If a satisfactory remedy is not found at an early date the yeild of good cocoons will gradually decrease on one hand and on the other hand the prospects of sericulture development would receive a setback.

NON-AVAILABILITY OF LAYINGS

Sericulturists generally buy layings (CBDFLs) from the government grainages and licenced seed-producers. The layings would be available adequately round the year except in the summer season. Some times the inadequacy, instability and poor quality of the CBDFLS exposes the sericulturists to some real trouble and lead to under- utilization of the available mulberry leaf. In such a situation sericulture becomes less economical. Some of the sample farmers express the opinion that disease free layings are very often affected by diseases.

The non-availablity of layings is one of the major problems faced by the sericulturists in Anantapur. The existing grainage facilities are inadequate and often fail to meet the demand. Normally, for every 405 hectares, there should be one grainage with a production capacity of ten lakh layings per year. Though 33,220 hectares (1993-94) of land are under mulberry cultivation in Anantapur, there are only twelve grainages, grossly inadequate to meet the legitimate demands of sericulturists. Therefore the sericulturists obtain layings, either from the nearby Karnataka state or from private grainages. To remedy the situation and to meet the growing demand, grainages for Local and Foreign races have also to be increased. Further the government has to take necessary steps to introduce bivoltine layings for commercial rearing in this district. However, in Kolar the problem of non-availability of layings is not so intense because there are fifteen grainages and a good number of licensed private persons who are also producing layings, in private grainages.

FLUCTUATIONS IN COCOON PRICES

Steady and speedy growth of any economic activity would not be possible without a stable and minimum economic price for its products. This statement is particularly applicable to the sericulture industry. There are violent fluctuations in the prices of its different products . These fluctuations are mainly due to a lack of guarantee of cocoon crop, wide variations in the quality of cocoons produced, absence of standardisation and quality control, poor and inadequate marketing facilities, and finally import of silk from other countries. During the rainy season, the quality of cocoons is likely to be poor though the yield may be higher. As a consequence the supply of cocoons in the market would be in excess causing a fall in the price of cocoons. During the survey conducted for the present study it is found that the price of cocoons fluctuated from Rs. 80 to Rs. 120 per kilogram of cocoons.

Moreover, sericulturists are forced to sell their cocoons at the prevailing price, because once the moth emerges out of the cocoon the cocoons become useless for reeling. As the sericulturists are forced to sell the cocoons within that time at the price prevailing in the market, their income suffers. In addition to this, the reelers also have the obligation to undertake the reeling work immediately.

To obviate these pressures stifling units can be established by the government or private enterpreneurs or both, so that the cocoons may be steam-stifled making the worm die in the cocoon. Thus, the cocoon can be preserved for a long time extending over months. Such of processing and preserving helps the sericulturists to realise better prices. The stifling technique faciitates the creation of buffer stocks of cocoons so as to maintain price stability. Therefore, steps have to be taken to develop the necessary infrastructure for stifling and building buffer stocks.

Further, the increase in the output of cocoons and raw silk should not be allowed to affect the price of cocoons through the manipulation by middle men. Fixation of cocoon and raw silk floor price is an essential ingredient of a package policy necessary to sustain the growth of sericulture.

SHORTAGE OF LABOUR

Mulberry cultivation does not require any technical knowledge or special skills. Hence, ordinary agricultural labourers with simple skills are enough for this activity. As mentioned earlier, in sericulture an established one acre mulberry garden provides 581 working days in a year. In spite of its employment potential, mulberry growers are suffering from shortage of labour. It is probably because mulberry can be cultivated even in small segments of landholdings. Hence, agricultural labourers who own less than one acre are also cultivating mulberry and silkworm rearing, thereby creating labour shortage for small as well as big farmers. Sericulture has been inducing not only small and marginal farmers but also agriculture labourers to take up self-employment. The labour shortage in cultivating mulberry can be overcome by introducing appropriate machinery in the mulberry fields on a co-operative basis. The State Government may be requested to help in their regard.

CLIMATIC DISTURBANCES

Cool climate throughout the year is a prerequisite for silkworm rearing, cocoon production and high renditta. Climatic disturbances upset cocoon production. In the tropical climate especially during summer sericulturists are advised to use air coolers, drip water on the rearing sheds, arrange the rearing rooms under the shade of big trees etc., to overcome the adverse effects of hot summer. The adoption of these new technieques though expensive marginally, reduces the adverse effects on the quantity and quality of cocoon and income per acre of the mulberry garden. High leaf yield from the mulberry garden raised on good soil which has good irrigation facilities may compensate the higher cost arising out of the extremes of summer climate. In any case climatic hazards do add to the drudgery of silkworm rearing necessitating greater care and attention on the part of sericulturists.

Other things being equal, the fact remain that a hot climate is a deterrent for sericulture entrepreneurship. As sericulture is a labour intensive enterprise, high wages that prevail in the agriucltural growth centres also constitute an obstacle for sericulture expansion. Big farmers, though they possess better capacity to face the climatic hazards, may not like the icreased managerial responsibility and time required to look after sericulture. In contrast, small farmers lack the resources to provide adequate facilities to counteract the hazards of severe heat during summer for silkworm rearing.

SHORTAGE OF SKILLED WORKERS

Crops other than sericulture may not need special skills and techniques. But it is the peculiar characteristic of sericulture activity that it require skilled workers, and the need for them is high during the period of silkworm rearing. As mentioned earlier, silk worm rearing is a very delicate activity which needs much care and attentin as silkworms at the infant stages look like ants. Tender mulberry leaves have to be cut into very fine pieces and spread on the silkworms. Generally silkworms take 25 to 28 days for obtaining complete growth and be ready to spin cocoons. During this period they undergo several changes and are very easily prone to diseases. Skilled labour is very much essential during this period to maintain timely feeding with required quantity and quality of leaves and also protect them against infecting diseases.

As sericulture is developing and progressing in both the study districts, steps should be taken to enhance the number of skilled workers and make them available. Training should be given to those interested in taking up sericulture and also the agricultural labour in the required skills at the district and block level centres. It, undoubtedly, would increase the yield of cocoons and reduce, unemployment among the rural population.

INADEQUATE FINANCE FOR INVESTMENT

The average sericulturist is either a small or a marginal farmer with limited means who requires financial assistance. As we know, the capital requirements for cultivating one acre of mulberry and silkworm rearing are definitely more than any other commercial or food crop . Therefore there is the need to assure adequate finance for making the required investment in sericulture. It is time that sericulturists have obtained from time to time institutional loans from commercial banks and cooperatives. But the amount received by them is far from adquate to meet their needs. Therefore, they have to borrow from other sources like friends, relatives and moneylenders at high rates of interest. Another problem regarding financial assistance is that the financial institutions are much too conservative to extend loans readily to sericulturists. Hence, for an orderly development of sericulture, institutional agencies should identify the special features and needs of sericulturists and the extent of finance required for fixed and working capital. Unless the institutional finance is streamlined the potentialities of sericulture may not be realised fully.

SHORTAGE OF REARING EQUIPMENT

The equipment required for silkworm rearing is of a special kind. While it is supremely important in sericulture activity, it can not be used for any other purpose. Since to possess the equipment needs considerable initial ivestment, only a few farmers can afford to procure it. Although banks and co-operatives are extending credit for this purpose, the amount sanctioned by them for obtining rearing equipment including shed is inadequate. Hence, majority of the silkworm rearers are hiring chandriakas and other equipment according to their necessity. Therefore, the State Government may help the sericulturists to supply them at subsidised rates.

INADEQUATE MARKET FACILITIES

Marketing facilities are inadequate in the Anantaput district where as in the Kolar district this problem is not felt. Absence of proper marketing facilities constitutes a major obstacle for the orderly and rapid development of sericulture. Seriuclturists of Anantapur are sending the cocoons mostly to the Karnataka markets. In the Karnataka regulated markets, the prevailing practice is that cocoons are purchasd by dealers in an open auction after a visual examination of the lots. Due regard to the quality of cocoons is not paid any where. The problems which sericulturists are facing in the Karnataka markets are mostly in the form of loss of sample cocoons supplied to the brokers, absence of proper weighing and unjustifiable deduction of certain percentage on the pretext that the produce is of inferior quality. Moreover, the market costs vary from place to place depending on distance and problems of transport.

At present there are six markets in the district of Anantapur, at Kadiri, Hindupur, Dharmavaram, Madakasira, Penukonda and Anantapur. Sericulturists from the other mandals of the district have to travel a long way to bring their cocoons to these markets facing the dificulties of transportaton, physical strain etc. Some of the rearers however take the cocoons to the Ramanagara, Vijayapura, Kanakapura and Sidlaghatta markets in Karnataka to get fair price. So, an efficient marketing organisation either through regulated markets or private dealers has to be set up giving it the top most priority. Appropriate infrastructure has to be built up with the creation of marketing facilities in the proximity of farmers so as to economise the cost of cocoon transportation. Moreover, efficient marketing organisation may help in arresting wide fluctuations in the price of cocoons by stabilising the market trends. Sufficient number of local markets may stimulate reeling acitivity and thereby promote silk industry in the district and thus generate more income and more employment.

SECURING EXTENSION OFFICERS ADVICE

As sericulture is not a routine type of activity, it requires constant improvement based on research results conducted by research agencies such as the Central Sericultural Research and Training Institute, Mysore and others. When the growth of sericulture is accelerated, more and more new sericulturists who joined the enterprise need scientific information, skill and knowledge regarding all aspects of it, including raising the mulberry garden and rearing silkworms. The field officers concerned are required to pay frequent visits to the sericulturists to guide them in their work, to check the spread of diseases as and when they are detected, to direct the farmers to cultivate better varieties of mulberry with porper application of manure, fertilizer and watering and to make them adopt corss breed races of silk worms and better methods of rearing. If the new sericulturists fail to utilise these extension services properly they would naturally fail in their new enterprise. Their failure, partial or full, may discourage other farmers and upset the process of innovation in introudcing the new enterprise. Speedy growth of sericulture postualates intensive extension services to facilitate healthy innovation.

In any case, speedy growth of sericulture, involving innovations on the part of the farmers will not be feasible without the establishment of a competent extension service department which has to enlighten and advice sericulturists. So the government has to build up a competent extension cadre to assist sericulture farmers.

CONCLUSION

Sericulture in the Anantapur and Kolar districts is leading towards prosperity, but the problems faced by the sericulturists hinder the progress of sericulture. New farmers may hesitate to take up this activity because of these problems. Hence, the government should be requested through the cooperatives to pay due attention to solve them. Otherwise, it will affect the growth of sericulture and those who are already engaged in sericluture activity may withdraw from it.

CHAPTER 9
Summary and Conclusions

Developing countries like India have still to rely very largely on the agriculture sector for economic development, which has occupied a vital place in the development strategies of India. However in recent years Indian agriculture is on the threshold of a stage of development characterised by a shift from static technology to modern technology.

In this context sericulture, with its vast potential for income and employment generation in rural areas, plays an important role in alleviating poverty. It is one of the crop enterprises which is identified as one of the most appropriate labour intensive cottage industries. This activity combines both agriculture and industry. Sericulture for example, has the credit of taking our tradition-bound agriculture into a modernised agriculture by intensive use of land, labour and capital.

Sericulture is best suited to a country like India where manpower and land resources are in surplus. It generates direct and indirect employment in various ways. By creating more employment opportunities in rural, urban and semi-urban areas, it not only arrests rural migration but also promotes a series of cottage and smallscale industries. 'Silk', the final product of the sericulture enterprise, has very good associations with the customs and traditions of the people living all over the world. Further, with improvement in the economic conditions of the people, the demand for silk is also increasing within the country and abroad.

Sericulture in India has a very long history. It is as old as Indian civilization. It has passed through a number of phases, rise and fall, in its long journey towards prosperity. During the last 20 years India has made tremendous progress in the production of mulberry silk for which there is an increasing international demand. Recently, sericulture in India has achieved enormous changes and progress by evolving suitable mulberry varieties, silkworm races and techniques suitable for the tropical climatic conditions. With the evolution and introduction of more productive silkworm races, productivity has increased and sericulture has become a highly remunerative activity.

More and more farmers in India have taken up sericulture activity and the industry which was once confined to five states only has spread to almost all the states of India. During the recent past, of all the states Karnataka has emerged as the leading producer of silk accounting for more than fifty per cent of the mulberry silk production in the country. This state is now regarded as the 'Silk Bowl of India'. Next to Karnataka, Andhra Pradesh though not a traditional state in silk production, occupies the second place in the country in producing mulberry raw silk.

Karnataka and Andhra Pradesh are showing enormous growth potential in sericulture, which in turn has improved the economic standards of their rural population. In this context analytical and empirical investigations carried out in these two states reveal several interesting factors which have been presented in the preceding chapters in detail. Here an attempt is made to put together all these observations, so as to present an overall view of the results emerging from the present study.

The main objective of the study is to examine the progress of sericulture in India in general and its position in Andhra Pradesh and Karnataka in particular. As it is not possible however to cover all the districts in these two states, to keep the study within manageable limits without affecting adversely the investigation, one leading district from each state i.e., Anantapur from Andhra Pradesh and Kolar from Karnataka are selected to analyse the economics of sericulture with particular reference to mulberry cultivation and silkworm rearing and to study the income and employment potentialities of sericulture activity.

To meet the above objectives, hypotheses were formulated to test whether (i) sericulture enables the small and marginal farmers (poor farmers) to become economically viable, (ii) mulberry cultivation and silkworm rearing provide more employment opportunities, and (iii) the initial investment for mulberry cultivation and silkworm rearing per acre is very high.

Historically China is the the motherland of silk. Chinese treasured the secret of silk for centuries together, but gradually it spread to Japan, Korea, India and other parts of the world. Now it is being practiced in 58 countries of the world. Sericulture enjoys a rapid growth mostly in South-East Asian countries and nearly 87 per cent of the world's raw silk production is from these countries only. Of these, China, stands first with 70 per cent of world's total raw silk production followed by India. Similarly Japan and Korea also have a long tradition as silk producing countries. Other countries like Bangladesh, Thailand, Nepal and Sri Lanka are steadily developing the sericulture industry. Some of the European countries where sericulture is being practiced are France, Italy, Switzerland etc. Brazil is the fifth largest producer and largest exporter of raw silk outside the Asian continent.

Sericulture in India is broadly classified into two distinct sectors namely mulberry and non-mulberry. Though India has the unique distinction of being the only country in the world which produces all the four commercially known varieties of silk, over 90 per cent of the silk produced in India is mulberry silk. Production of mulberry raw silk in the country is mainly confined to the states of Karnataka, Andhra Pradesh, West Bengal, Tamil Nadu and Jammu and Kashmir. With the liberal financial allocation in the Five Year Plans and the contribution of the Sericulture Department through its research, sericulture has proved to be an effective tool for eradication of rural poverty and unemployment. In India sericulture has been employing 60.30 lakh persons in about 59,528 sericulture villages all over the country. The area under the mulberry crop which was only 1,70,000 hectares in 1980-81 increased to 3,41,200 hectares in 1993-94.

The promotional agencies have also played an essential and dynamic role in the development of sericulture. The Central Silk Board, Bangalore and the Central Sericultural Research and Training Institute, Mysore, have been helping by providing suitable technologies, evolving new varieties of mulberry and silkworms. Further, the biggest impetus in recent times to the development of Indian sericulture is the implementation of National Sericulture Project (NSP) with Rs.555/- Crores financial assistance from the World Bank and the Swiss Development Co-operation. With increased financial and technical assistance, from these agencies sericulture has been extended to new areas and non-sericulture states. Presently it is being practiced in 23 states of India, of which only five are traditional states and the remaining 18 are non-traditional states.

Karnataka, a traditional silk state, has been called the 'Silk Bowl of India' as it produces more than fifty per cent of the total raw silk production in the country. Andhra Pradesh comes next in this regard. The general climatic conditions in these two states are quite favourable for the growth of sericulture. Though sericulture is a new enterprise to the farmers in Andhra Pradesh, the growth rates (1980-81 to 1993-94) in the area under mulberry, cocoon production and raw silk production are significantly more when compared to those of Karnataka . In Andhra Pradesh during the period from 1980-81 to 1993-94, covering 14 years, the area under mulberry increased from 18.90 to 90.77 thousand hectares, reeling cocoon production increased from 10.98 to 24.51 thousand tonnes and the raw silk production increased from 0.08 to 2.86 thousand tonnes. But infrastructural facilities have not grown commensurately to meet the increasing demands of this industry in the state. Therefore majority of the farmers of this state carry their cocoons to the Karnataka silk cocoon markets.

In Karnataka, which ranks first in raw silk production, the area under mulberry has increased from 116.19 to 160.84 thousand hectares, reeling cocoons production increased from 38.01 to 70.21 thousand tonnes and raw silk production from 2.88 to 8.25 thousand tonnes, (1980-81 - 1993-94).

The National Sericulture Project and many other projects and programmes which are under implementation in these two states are also responsible for the commendable growth of sericulture.

Both the study districts, Anatapur and Kolar, are adjacent to each other and are drought prone districts. They have the unique distinction of raising mulberry under irrigated conditions. In spite of recurring droughts there is enormous growth of mulberry during the last 15 years, as mulberry is drought resistant.

Sericulture has become a most promising activity for the farmers in these two districts. In Anantapur the area under mulberry increased from 11.78 to 33.22 thousand hectares and cocoon production increased from 7.72 to 11.47 thousand tonnes during period from 1980-81 to 1993-94. Raw silk production increased from 0.19 to 0.31 thousand tonnes during the same period. In Kolar where the growth of sericulture is very encouraging, the area under mulberry increased from 16.36 to 26.85 thousand hectares, cocoon production from 9.61 to 20.00 thousand tonnes and raw silk production increased from 0.73 to 2.35 thousand tonnes during the period from 1980-81 to 1993-94. The infrastructural facilities available in these two districts have also helped the fast growth of sericulture. However, there is a need to improve the infrastructural facilities in Anantapur to meet the growing demand of sericulture.

In Anantapur the concentration of mulberry is more in the Hindupur, Madakasira and Kadiri mandals and hence these three mandals were selected for the sample survey. Similarly, sericulture activity is more traditional and popular in the Kolar, Chintamani and Sidlaghatta taluks of the Kolar district. Therefore, these three taluks were selected for the sample survey. To make the sample farmers, for the study, three villages were selected from Anantapur drawing one from each mandal i.e., Rachapalle from Hindupur, Papasani palle from Madakasira, Motukupalle from Kadiri.

Similarly three villages were selected from Kolar drawing one from each taluk Sugutur from Kolar, Doddaganjur from Chintamani and Sonnanahalli village from Sidlaghatta . These six sample villages are located within a 10 to 15 Kms. radius from their respective mandal/taluk headquarters. The concentration of mulberry in each village is high i.e., around 200 to 300 acres. Mulberry is cultivated in these villages under irrigated conditions and the source for irrigation are bore wells and dug wells. There are around 150 to 200 sericulturists in each village. The sample farmers were drawn at random at the rate of 30 farmers (20 sericulturists and 10 were non-sericulturists) from each village making the total sample 180. Thus, the study covered 120 sample sericulturists (60 from each district) and 60 non-sericulturists (30 from each district).

Of the total 180 sample farmers 44 are marginal farmers with an average landholding of 1.68 acres, 65 are small farmers with average landholding of 3.99 acres and the remaining 71 are big farmers with an average landholding of 12.01 acres. The average landholding per household is 6.48 acres in the Anantapur district, and 5.93 acres in the Kolar district. Of the total sample, 20 per cent belong to the SC/ST, 17 per cent belong to the BC and the remaining 63 per cent belong to the OC. Regarding the literacy levels of the farmers 49 per cent are illiterates, 43 per cent have school education and 8 per cent have collegiate education. Out of the 6.48 acres average landholding per household in Anantapur, only 1.51 acres (23 per cent) has irrigation facility whereas in Kolar the average area under irrigation is 1.89 acres (32 per cent) as against the 5.93 acres of average landholding per household.

In the Kolar district bore-wells are the major source of irrigation, whereas in the Anantapur district the sources of irrigation are bore wells and few dug-wells. The command area under bore wells is naturally greater than the area under dug wells and hence, area under irrigation per household was more in Kolar than in Anantapur.

The cropping pattern reveals that groundnut in Anantapur and ragi in Kolar dominate the other crops. The productivity levels of paddy, groundnut and ragi are higher in Kolar than the levels in Anantapur. The main source of income for the sample households is agriculture though there is some income form the milch cattle. The agricultural income per household is of the order of 81 per cent in Anantapur and 69 per cent in Kolar. The remaining 19 per cent income in Anantapur and 31 per cent in Kolar comes from milch cattle. The income per household through milch cattle is far higher in Kolar and supports its reputation for 'silk and milk' in Karnataka.

The comparative study of the socio-economic conditions of sericulturists and non-sreriсulturists reveals that the percentage of illiterates among non-sericulturists is higher (67 per cent) than among sericulturists (40 per cent). The average of holdings per household is much higher among sericulturists (around 7.21 acres) when compared to 4.22 acres among the non-sericulturists. Out of the total landholdings nearly 10 per cent of the land held by the non-sericultirists and 32 per cent of the land belonging to the sericulturists have irrigation facilities. Among sericulturists because of their better financial position, the average size of their land holdings is bigger and the area brought under irrigation is also greater. The unit cost of construction of a bore well would be around Rs.60,000/- and most of the non-sericulturists are not in a position to invest such huge amount to bring their land under irrigation. The cropping pattern reveals that the sericulturists have given priority to mulberry over paddy, groundnut and ragi in both the districts under study. Among the non-sericulturists in Ananatapur give priority to groundnut and to ragi by their counterparts in Kolar. It is evident that because of the higher returns that mulberry promises the concentration is on sericulture.

Sericulture also makes its contribution to the dairy enterprise and adds to the income of the sericulturists. Its by-products form good cattle feed and hence, the sericulturists are able to maintain more number of milch cattle than the non-sericulturists. This is evident from the fact that the per household income from milch cattle is far higher among the sericulturists than that of the non-sericulturists in both the study districts. Thus, the farmers who have been practicing sericulture are better placed economically.

Although these two districts are subject to frequent droughts, the area under mulberry in them has increased year after year because the plant is drought resistant and gives more income in shorter intervals. A number of variables which focus on the structure of costs and returns among the sample households with different sizes of mulberry landholdings have been examined.

The Majority of the sample households have adopted sericulture far more than 10 years in the Kolar district and between five to ten years in the Anantapur district. Most of the mulberry landholdings are below one acre and one to two acres. Mulberry land holdings, which are more than two acres are only a few. Hence, one hectare becomes a large unit and therefore one acre is adopted as the unit for analysing the variables that influence the sericulture practice. The distribution of mulberry landholdings in Anantapur shows that 39 sericulturists have mulberry gardens of less than one acre area, 18 have between one to two acres and the remaining three cultivate mulberry in a garden of more than two acres each. In Kolar 31 sericulturists have mulberry garden of less than an acre, each 22 sericulturists have garden varying between one and two acres, and 7 farmers have gardens of more than two acres. The average size of mulberry landholdings is 1.18 acres in Ananatapur and 1.48 acres in Kolar. Thus, the average size of mulberry landholdings is slightly higher in Kolar.

As Mulberry is a perennial crop, the initial establishment of the garden is of crucial importance and no compromises can be made on the initial establishment. The profitability of sericulture largely depends on the production of mulberry leaf and its conversion into quality cocoons at economic costs. The garden once established yields for about 12 to 15 years. The mulberry plantation is made with cuttings according to the row system which is popularly called the 'Kolar System'. Both the Local and Kanva-2 (M5) are the varieties of mulberry used by the sample sericulturists. The estimated cost of initial establishment of one acre irrigated mulberry garden is around Rs.5,165/- in Anantapur and Rs.4,550/- in Kolar. The management cost of this garden during the establishment year is estimated to be Rs.4,890/- in Anantapur and Rs.5,250/- in Kolar. Thus, the total estimated cost of the establishment and management of one acre irrigated mulberry garden during the first year works out to be Rs.10,055/- and Rs. 9,800/- in the Ananatapur and Kolar districts respectively.

Once the mulberry garden is established it gives its yield over a long period. The recurring expenditure involved in maintenance of one acre of irrigated mulberry garden from the second year onwards is estimated as Rs.10,680/- in Anantapur and Rs.9,975/- in Kolar. With the involvement of family labour this expenditure is reduced significantly and is worked out to be Rs.6,170/- and Rs.4,982/- in Anantapur and Kolar districts respectively. Thus, with the involvement of family labour the recurring expenditure of the sample households in the total estimated cost is only 58 per cent in Anantapur and 50 per cent in Kolar.

The mulberry leaf production per acre per annum is of the order of 7,862 and 9,242 Kgs in Anantapur and Kolar making the production cost of one Kg. of mulberry leaf Rs.0.78 and Rs.0.54 in the two districts respectively. It is interesting to note that the sericulturists of Kolar are able to get more leaf yield per unit area with less cost. This is because of their rich experience in this activity.

Silkworm rearing which is a cottage activity, requires a lot of managerial attention and technical skills. It has to be done on protected premises to avoid infection and disturbances. In both the Anantapur and Kolar districts around 62 per cent of the sericulturists undertake the rearing activity in their dwelling houses, only around 13 per cent have separate rearing sheds in the farm, and 25 per cent of sericulturists have a separate room for rearing in their dwelling houses. Rearing in dwelling houses results frequently in infection to the silkworms. Therefore, a separate and well-protected shed is essential to improve the quality and productivity of silk cocoons.

The cost of investment on a rearing shed is around Rs.25,000/- and on rearing equipment is Rs.3,423/- (for rearing 200-250 dfls). The quantity of silkworms reared depends on the size of the mulberry garden. The annual quantity of disease-free layings reared per acre is around 1,000 and 1,100 in Anantapur and Kolar respectively. Generally the sericulturists rear four to five crops in a year. Among the sample sericulturists in Anantapur the average number of rearing in a year is six. Now-a-days the sericulturists with large area under mulberry, are dividing their garden into two or three parts and undertake 8 to 10 rearings in a year. On an avenge, the sample seiculturists rear 200 to 250 dfls per acre per crop. The cost estimated to rear 200 to 250 dfls is Rs.2,663 and Rs.2,175 in the Anantapur and Kolar districts respectively. As stated earlier, on an average four rearings in Anantapur and six rearings in Kolar are undertaken in a year. Thus, the annual expenditure on rearing per one acre of mulberry works out to be Rs.14,572/- in Anantapur and 16,970/- in Kolar district. This expenditure is inclusive of depreciation, on the rearing shed and non-recurring expenditure on the rearing equipment.

The rearing activity is labour intensive and accordingly the estimated expenditure on labour component is also more when compared to the other inputs involved in this activity. So the involvement of family labour has reduced the cost of rearing substantially and the average expenditure incurred by the sample households on rearing per acre in a year is Rs.9281/ - and Rs. 9,685 in Anantapur and Kolar, respectively. In the total estimated costs of Rs.14,572/- in Anantapur and Rs.16,970/- in Kolar district, the involvement of family labour has reduced the annual expenditure on rearing by 36 and 43 per cent in the districts in their respectively order.

Therefore, the involvement of family labour is a decisive factor in both mulberry cultivation and silkworm rearing to make the enterprise economical. The family labour involvement has reduced the mulberry cultivation cost on an average, by 46 per cent and silkworm rearing cost by 40 per cent.

The economics of sericulture depends on the successful rearing of silkworms and the cocoon yield. Failures of crops do occur due to diseases infecting silkworm resulting in crop losses. The average failure rate among the sample households works to be 15 per cent in Anantapur and 22 per cent in Kolar, which does not seem to be alarming. To derive one kilogram of cocoons, the leaf used should be around 30 Kg. Anything more than this reflects the inefficient use of the leaf. The rearers in the study area have shown high efficiency. Particularly the rearers of Anantapur have shown better efficiency with 20 Kg per 1Kg. of cocoons than those in Kolar who use 23Kg. of leaf for a like yield. The average cocoon yield is more or less the same in both districts and it works out to be around 47 Kg per 100 dfls harvested.

The success of sericulture enterprise depends on the level of profits it brings. In the expenditure on mulberry cultivation and silkworm rearing, the cost involved in rearing silkworms is 61 to 66 per cent of it, as the rearing activity demands utmost care and skill on the part of the laboures who attend to it. The average net returns per year from one acre of mulberry garden are of the order of Rs.23,150/- in Anantapur and Rs.25,700/- in Kolar. The average annual income per household through sericulture works out to be Rs.27,281/- in Anantapur Rs.37,850/- in Kolar. This difference is due to the higher average area under mulberry per household and better operational skills in Kolar. (i.e., 1.18 acres in Anantapur and 1.48 acres in Kolar).

The sericulture respondents also cultivate other crops like paddy, groundnut, ragi, small millets and vegetables. Owing to continuous research and improvement in recent years, sericulture has established its economic superiority over the other principal crops with a satisfactory cost-benefit ratio. The input costs of sericulture are very high (Rs.15,452/- in Anantapur and Rs.14,666/- in Kolar), when compared with those of other crops in the study area.

The input cost in sericulture is two and half times more than that of paddy and groundnut and five times that of ragi. This higher level of input costs is associated with higher returns in sericulture. Though paddy, groundnut and ragi are taken up as double crops, the net returns from sericulture per annum per acre are two and half times more than that of paddy and groundnut, and nearly twenty three times that of ragi. Sericulture thus makes economic use of valuable resources like land and water and gives higher returns than paddy, groundnut and ragi, thereby improving the economic conditions of the small and marginal farmers who take to it.

To identify the variables which have their impact on the production of mulberry leaves and cocoons, the Principal Component Analysis was employed. It was found that three variables, namely area under irrigation, area under mulberry crop and size of landholdings largely account for the total variation in both the study districts. To study the influence of each factor in mulberry leaf production Cobb-Douglas production function was used and it was found that the co-efficient of landholdings in the Anantapur and Kolar districts are significant at one per cent level indicating the positive influence on the production of mulberry leaves. The same Cobb-Douglas production function was also used to study the influence of each variable on cocoon production. It was found that the co-efficients of the number of disease-free layings (dfls) in Anantapur and in Kolar are significant at 1 per cent level.

Employment opportunities in sericulture can be categorised under two heads. Those opportunities relating to mulberry cultivation and silkworm rearing which are agricultural in nature and which are undertaken in rural areas, and secondly those opportunities relating to silk reeling, twisting, weaving and marketing which are undertaken mostly in semi-urban and urban areas. Here the employment opportunities relating to the first category are studied and compared with other crops.

The establishment and management of one acre of mulberry garden during the first year creates 631 man days of which 269 are in the establishment of mulberry garden, 162 in management and 200 in silkworm rearing. Employment opportunities generated from the second year onwards on one acre irrigated mulberry garden are 586 man days per year, of which 262 in the management of mulberry garden and 324 in silkworm rearing. Thus, silkworm rearing has more employment potentiality than mulberry cultivation. Further, the study reveals that the generation of employment opportunities per household are more in the group of sericulturits who own more than two acres of land, and it is worked out to be 1,546 and 2,358 man days per household in the Anantapur and Kolar districts respectively. The generation of man days in each class of mulberry holdings as well as the pooled average per household is more in Kolar. And that is due to more number of rearings undertaken by the sericulturists in Kolar and also due to a higher average size of the mulberry holdings.

A comparative study of the employment opportunities generated from sericulture and other principal crops reveals that the generation of man days on one acre of mulberry garden is three times more than that of paddy, nearly four times more than that of groundnut and five times more than that of ragi in the Anantapur district where as in the Kolar district it is four and half times more than that of paddy and groundnut and five times that of ragi. On an average 586, 153, 135 and 110 man days are generated from one acre cultivation of mulberry, paddy, groundnut and ragi respectively under irrigated conditions per year. The calculated 't' values are significant at one per cent level indicating that there is a significant increase in employment generation due to sericulture in both the districts. The test results also indicate that employment generation in Kolar is more when compared to Anantapur.

Sericulture also creates gainful employment to women and aged people at homes with minimum risk. More over, a number of indirect employment opportunities can also be created in the ancillary sectors of sericulture like manufacturing rearing appliances, processing and spinning silk waste and in the extraction of pupae oil and pupae meal from the by-products of sericulture.

Thus, the analysis clearly establishes the importance of sericulture over other crops in the generation of fresh employment opportunities in rural areas. Sericulture provides higher returns and gainful employment throughout the year relieving the farmers, more particularly small and marginal farmers, from the clutches of disguised and seasonal unemployment. This enterprise makes an ideal use of manpower, water and land resources which are the critical inputs in drought prone areas where, the strategy of farm development should be the economic use of these inputs with a view to maximise returns.

In a growing enterprise like sericulture, problems are bound to be many and sometimes unavoidable. But these problems should not impede the progress of this enterprise towards long range goals. The problems faced by sericulturists mainly relate to the diseases which infact silkworms, availability of layings, shortage of labour, insufficient financial support from the government agencies, climatic hazards, wide fluctuations in cocoon prices, and also to some extent inadequacy of extension services.

Of these the most important are the diseases infecting silkworms, inadequate availability of layings, fluctuations in cocoon prices, shortage of skilled labour and insufficient financial support for investment on rearing shed and equipment.

To overcome these problems and to make this enterprise more attractive the following suggestions are made.

There is need to introduce new varieties of mulberry like S_{13}, S_{54}, and S_{36}, which have been found viable with regard to the quality and quantity of leaf by the research institutions.

The sericulturists must be encouraged to rear bivoltire variety of silkworm rather than multivoltine silkworms to obtain better income.

The average cocoon yield of 47 Kg per 100 dfls can be further raised by following proper techniques of silkworm rearing.

More farmers may be motivated to take up sericulture on a largescale through co-operative farming.

Inadequate irrigation facilities are hindering the growth of sericulture. Hence, irrigation facilities have to be bettered and enhanced by giving incentives to the farmers for digging wells.

Some of the farmers who want to shift to sericulture are scared of the diseases inffecting silkworms. Hence, research organisations should provide such potential farmers with the necessary technical knowledge and suggest the methods and means to overcome this problem.

To meet the growing demand for layings, adequate number of grainages should be opened and care must be taken to supply good quality layings through them.

Stifling units can be established both by the government and private entrepreneurs to preserve the cocoons for longer periods to overcome the problem of fluctuations in cocoon prices.

Labour shortage in the cultivation of mulberry can be filled by introducing appropriate machinery on a co-operative basis which can be used by the sericulturists.

To reduce the intensity of summer, air coolers, dripping of water in sheds, arranging rearing rooms under tree shades, hanging wet gunny bags on the doors and windows of rearing rooms and the use of sand beds and like devices may be employed.

Provision of training facilities to the labourers can solve the problem of shortage of skilled workers.

The unit cost of Rs.19,800/- given for establishing one acre of mulberry garden to the sericulturists in Andhra Pradesh is quite inadequate. Hence, it should be enhanced according to the cost of inputs.

Necessary equipment for rearing silkworms may be supplied to the sericulturists at subsidised rates.

Improvement and enhancement of infrastuctural facilities in Andhra Pradesh will definitely promote the further growth of sericulture in the state.

Extension services play a key role in the expansion of the sericulture industry. Therefore, there is need to extend these services much further, particularly to the remote and inaccessible areas.

Crop insurance may be considered with the subsidised premium rates, to give the sericulturists a sense of security. Mulberry silk, has a very strong and expanding domestic and export market. It brings money from the richer to the poorer sections as well as helps in earning valuable foreign exchange for the country. In Andhra Pradesh and Karnataka the scope for the development of sericulture is clear-cut in two directions. One relates to the expansion of sericulture in new areas and the other to modernising the industry in the existing areas. There is a lot of scope for the development of sericulture in the Southern Regions of Andhra Pradesh which needs to be given special attention in view of their climatic and natural advantages. Further, efforts have to be simultaneously directed towards the improvement of mulberry varieties as well as silkworm races. A suitable variety of mulberry which is adoptable to the local environment and resistant to high temperature has to be developed.

Sericulture activity in India is basically individual farmer-oriented. Hence, a comprehensive programme to cover the entire area in the districts with the new practices should be drawn up. Qualified and experienced staff should be deployed for extension work. Subsidies and incentives should be given to progressive farmers.

In the methods of silkworm rearing, the most advantageous and economical technique is the system of institutional rearing of young age silkworms (chawkie). In places where there are large groups of rearers, there is scope for organisation of co-operative chawkie rearing-cum-reeling units. Apart from having separate rearing building and proper equipment for the maintenance of humidity and temperature, the farmers should also maintain hygienic conditions in the rearing areas. Training and extension centrers have to be established at mandal/taluk levels to educate the farmers in reducing the incidence of diseases to silkworms and to get higher cocoon crops.

Further, as a predominant sector of rural development stability is the vital need of sericulture industry. Hence, efforts should be made to put the fortunes of this labour intensive activity on sound lines by establishing regional organisations for stabilising the silk prices. This step definitely ensures a fair price to the primary producer namely the silkworm rearer, at a level that would ensure stability and further promotion of mulberry cultivation and silkworm rearing.

Bibliography

BOOKS

Abdul Aziz & Hanumappa, H.G., : **Silk Industry - Problems and Prospects,** Ashish Publishing House, New Delhi, 1985.

Acharya, J. : **Sericulture and Development, Development Sociology Series** - 1 Indian Publishers Distributors, Delhi, 1993.

Agarwal, S.K., & Others :**Agricultural Economics and Co-operation,** S.Chand & Co Ltd., New Delhi, 1970.

Aruga Hisao : **Principles of Sericulture,** Oxford & IBH Publishing Co., (Pvt) Ltd., New Delhi, 1994.

Bansal, P.c. : **Agricultural Problems of India,** Oxford & IBH Publishing Company (Pvt) Ltd., New Delhi, 1981.

.... : **Agriculture Situation in India - A Guide,** Oxford & IBH Publishing Company, New Delhi, 1984.

Benerjee, P.K. : **Indian Agricultural Economy - Financing Small Farmers,** Chetana Publications, New Delhi, 1977.

Benjamin, R.E., Ltd., & Others : **Economics of Agriculture,** S.Chand & Co. New Delhi, 1989.

Bepin, Behari. : **Rural Industrialisation in India,** Vikas Publishing House, Ghaziabad, 1974.

Central Silk Board : **Silk Man's Companion,** Bangalore, 1989.

.... : **Silk Man's Companion,** Bangalore, 1990.

.... : **Silk Man's Companion**, Bangalore, 1992.

Central Sericultural : **Achievements of CSRTI**, Mysore.Research & Training Institute

.........., : **Central Sericultural Research & Training Institute, Mysore - Its Organisational Set Up**, 1981.

Charsley, S.R. : **Culture & Sericulture** Academic Press, New York, 1982.

Chowdhury, M.R. : **Indian Industries Development & Location**, Fourth Edition, Oxford & IBH Publishing Co., (Pvt) Ltd., New Delhi, 1970.

Dhar, P.N., & Lydall, H.F., : **The Role of Small Enterprises in Indian Economic Development**, Asia Publishing House Bombay, 1961.

F.A.O. : **Manual of Sericulture - I, Mulberry Cultivation**, Oxford & IBH Publishing Company, (Pvt) Ltd., New Delhi, 1993.

..........., : **Manual of Sericulture - 2 Silk Worm Rearing**, Oxford & IBH Publishing Company, (Pvt) Ltd., New Delhi, 1993.

..........., : **Sericulture Training Manual**, Oxford & IBH Publishing Company., New Delhi, 1994.

Ganga, G., & Sulochana Chetty,J., : **An Introduction to Sericulture**, Oxford & IBH Publishing Copmany, (Pvt) Ltd., New Delhi. 1991.

Ghosh, Alok : **Indian Economy, 1988-89. It's Nature and Problems**, The World Press Private Ltd., Calcutta, 1988.

Ghosh, C.C., : **Silk Production and Weaving in India**, CSIR, New Delhi, 1949.

Gopalachar, A.r.S. : **Three Decades of Sericulture Progress**, CSB, Bangalore, 1978.

Giri,V.V. : **Jobs for Millions**, 1970.

Gyan Chand : **Population in Perspective**, 1972.

Hanumappa, H.G. : **Sericulture for Rural Development**, Himalaya Publishing House, Bombay, 1986.

.........., : **Sericulture Society and Economy**, Himalaya Publishing House, Bombay, 1993.

Harpal Singh, Y. : **Project for the Development of Sericulture**, National Institute of Bank Management, Bombay.

Iqbal A.Badar : **Financing of Agro-Industrial Development in India**, Kmar Publications, Aligarh, 1979.

Jagdananad Jha : **Khadi and Village Industries in Economic Development**, Deep & Deep Publications, New Delhi, 1990.

Jha, D.N. : **Planning and Agricultural Development**, Vikas Publications, New Delhi, 1974.

Jolly, M.S. : **Appropriate Sericulture Techniques**, Director, International Centre for Training & Research in Tropical Sericulture, Mysore, 1987.

Jolly, M.S., Chowdhury: **Non-Mulberry Sericulture in India**, S.N., & Sen, S.K. Central Silk Board, Bombay, 1975.

Krishna Swamy, S. : **Mulberry Cultivation in South India**, CSR & TI, CSB, Govt. of India, Ministry of Textiles, Bangalore, 1986.

.........., : **Improved Method of Rearing Young Age (Chawkie) Silk Worms**, CSR, TI, CSB, Govt.of India, Ministry of Textiles, Bangalore, 1986.

.... , : **New Technology of Silk Worm Rearing,** CSR&TI, CSB, Govt.of India, Ministry of Textiles, Bangalore, 1986.

.... , : **Economics of Sericulture Under Irrigated Conditions,** CSR & TI, CSB, Govt.of India, Ministry of Textiles, Bangalore, 1986.

Koshy, K.D. : **Silk Exports and Development,** Ashish Publishing House, New Delhi, 1993.

Momoria, C.B. : **Agricultural Problems of India,** Kitab mahal, Allahabad, 1984.

Michael P. Todaro : **Economic Development in the Third World,** Orient Longman, Hyderabad, 1991.

Misra, R.P. : **Rural Industrialisation in Third World Countries,** Sterling Publishers, (Pvt) Ltd., New Delhi,1985.

Mukerji, N.G. : **Hand Book of Sericulture,** Bengal Secretariat Book Department, Calcutta, 1906.

Myrdal Gunnar : **Asian Drama - An Enquiry into the Poverty of Nations,** The Twentieth Century Fund, Inc., Vol.-11, London, 1968.

Mysore Silk Association: **Silk Worm Rearing and Diseases of Silk Worms,** Bangalore, 1956.

Nanavathy M.Mahesh : **Silk- Production, Processing and marketing,** Wiley Eastern Limited, New Delhi, 1990.

Narayana, D.L. : **Economics of Sericulture in Rayalaseema,** Technical Cell, S.V.University, Tirupati, 1979.

.... , : **Employment and Economic Growth,** Madhurai Kamaraj University, Madhurai, 1970.

.... , : **Entrepreneurship and Agricultural Development,** Indian Institute for Asian Studies, Bombay, 1966.

Raj Purohit, A.R., & Govinda Raj, K.V. : **Employment and Income in Sericulture,** Shiny Publications, Bangalore, 1981.

Rao, R.V. : **Rural Industrialisation in India,** Classical Publishers, New Delhi, 1973.

Ramakrishna Sharma : **Industrial Development of Andhra Pradesh,** Himalaya Publishing House, Bombay, 1982.

Ramana, D.v. : **Economics of Sericulture and Silk Industry in India,** Deep & Deep Publications, New Delhi, 1987.

Ramanadham, V.V. : **Economics of Andhra Pradesh,** Asia Publishing House, Bombay, 1959.

Rowley, R.c. : **Economics of Silk Industry,** P.S.King & Song Ltd., London, 1919.

Sadhak, H. : **Industrial Development in Backward Regions in India,** Chugh Publications, Allahabad, 1986.

Santha Raj Kumar S.B.: **Silk Handloom Industry in Andhra Pradesh,** Sonali Printers, Pune, 1986.

Sinha Sanjay : **The Development of Indian Silk,** Oxford & IBH Publishing Company (Pvt) Ltd., New Delhi, 1986.

Thammanna. N., & : **Hand Book of Silk Technology,** Wiley Sonwalkar Eastern Limited, New Delhi, 1993.

Tazima, Y. : **Sericulture Industry in India,** CSB, Bombay.

Ullal, S.R., & Narasimhanna, M.N. : **Hand Book of Practical Sericulture,** CSB, Bombay,1978.

Venkata Narasaiah, P. : **Sericulture in India,** Ashish Publishing House, New Delhi, 1992.

Venkaiah, V. : **Impact of Agrio - Based Industries on Rural Economy,** Himalaya Publishing House, Bombay, 1987.

REPORTS

Central Silk Board : **Souvenir, International Congress on Trophical Sericulture Practices, (18-23), February,** Bangalore, 1985.

..........., : **Proceedings of the International Congress on Tropical Sericulture Practices, (18-23), February,** Bangalore, 1988.

..........., : **Statistical Biennial, 1986.**

..........., : **Statistical Biennial, 1988.**

..........., : **Statistical Biennial, 1990.**

..........., : **Statistical Biennial, 1992.**

..........., : **Sericulture Development Programme During VIII Plan (1992-97) and Annual Plan 1992-93 Under Central Sector,** Bangalore, 1991.

..........., : **The NSP News letter, A Monthly Publication of the National Sericulture Project,** Vol.-1, Nos.1-12. Bangalore, 1991.

..........., : **A Note on Present Status of Indian Sericulture, its Silk Production and Exports,** Bangalore, 1987-88.

CSR & TI : **Seridoc, A Quanterly Document of Sericultural** Research and Training Institute: Mysore.

Centre for Planning & Development Studies : **Evaluation of the Impact of Drought Prone Areas Programme in Anantapur District,**1987, Anantapur (unpublished report).

Centre for Rayalaseema Development Studies : **Development of Sericulture and its impact on Cropping pattern in Rayalaseema districts,** S.K.University, Anantapur (unpublished report)

T. Chandra Reddy : **Impact of Sericulture Industry on Income and Employment in Rural Area of Chittoor district of Andhra Pradesh** (unpublished Ph. D. thesis submitted to U.A.S., Bangalore, 1987.)

Government of Andhra Pradesh : **Planning and Development of Backward Regions - A Case Study in Rayalaseema,** Vol.1.

.........., : **Annual Administration Reports 1980-81 to** 1993-94, Dept. of Sericulture, Hyderabad.

.........., : **A Note on Implementation of National Sericulture Project in A.P. (Summary and** Progress) - Dept. of Sericulture, Hyderabad, 1991.

.........., : **Hand Book of Statistics, Anantapur District,** CPO, Anantapur, 1991-93.

.........., : **Pattu Sravanthi,** A Quarterly Document of the Dept. of Sericul-ture, Hyderabad.

.........., : **Action Plan for Women in Sericulture in A.P.,** Dept. of Sericulture, Hyderabad.

.........., : **Andhra Pradesh State Gazetteer,** Hyderabad.

.........., : **Sericulture Development Under DPAP;** Anantapur District, Dept.of Sericul-ture, Anantapur, 1987.

.........., : **New Horizons for Sericulture in Andhra** Pradesh, Dept. of Textiles and Handlooms, Hyderabad, 1976.

.........., : **Brief Note on the Development of Sericulture** in Anantapur District, Dept. of Sericulture, Anantapur, 1994.

.... , : **Status Paper on Sericulture in Andhra Pradesh,** Orientation Programme in Sericulture for Bankers from A.P. (24th June - 1st July) 1991. Dept. of Sericulture, Hyderabad, 1991.

.... , : **Crop and Season Reports,** Directorate of Economics and Statistics, Hyderabad, 1980-81 to 1993-94.

.... , : **Master Plan for Sericulture Development in Andhra Pradesh,** (1981-82 to 1985-86).

.... , : **Block Plans 1980-81, Anantapur District,** Intensive Development of Block Under the Integrated Rural Development Programme.

.... , : **Anantapur District Gazetteer,** Hyderabad.

.... , : **Credit Plan for Sericulture in Andhra Pradesh,** Department of Handlooms and Textiles, 1978.

.... , : **District Plans Under Drought Prone Areas Programme, Integrated Rural Development Programme and District Rural Development Agency Programme,** Anantapur District.

.... , : **A brief note on World Bank Assistance Programmes,** Directorate of Sericulture, Hyderabad, 1988.

.... , : **A brief note on Sericulture Project Under Indo-Swiss Programmes,** Dept.of Sericulture, Anantapur, 1987.

.... , : **Report on Swiss Assistance Programmes,** Dept. of Sericulture, Anantapur, 1989.

.... , : **Report on S.C. Action Plan, 1984-85 to 1988-89,** Dept. of Sericulture, Anantapur.

..........., : Industrial Profile on Anantapur District, General Manager, District Industries Cen Anantapur, 1987.

..........., : Progress Reports, 1980-81 to 1993-94, Dept. of Sericulture, Anantapur.

Government of India : Drought Prone Areas Programme, Ministry of Agriculture, Department of Rural Development, New Delhi, 1978.

..........., : First Five Years Plan - A Draft Out Line, Planning Commission, New Delhi, 1957.

..........., : Eighth Five Years Plan, (1992-97), Vol.I, & Vol.II Planning Commission, New Delhi, 1991.

..........., : Census of India, 1991, Series I and 2, District Census Hand Book, Anantapur District Parts XIII-A & B, Village, Town Directory, Village, Town Wise Primary Abstract.

..........., : Report on financing of Small Scale Industries, State Bank of India Study Team,

Government of Karnataka : Karnataka State Gazetteer, Bangalore.

..........., : Brouchure on Land Use Statistics in Karnataka, 1980-81 to 1991-92, Directorate of Economics & Statistics, Bangalore, 1993.

..........., : Annual Administration Reports, Department of Sericulture, Bangalore, 1980-81 to 1993-94.

..........., : Note on Kolar District Statistics, C.P.O. Kolar 1990-91.

..........., : District Socio-Economic Indicators, Directorate of Economics and Statistics, Bangalore, 1993.

.... , : Mysore State Gazetteer, Bangalore.

.... , : Kolar District, The Director, Printing Stationery and Publications, Bangalore.

.... , : Draft VIII Five Years Plan, Depart-ment of Sericulture, Bangalore, 1990.

.... , : NSP, Agenda & Notes for Review, Dept. of Sericulture, Bangalore, 1990.

.... , : NSP, (Karnataka Sericulture Project-Phase-II), Department of Sericulture, Banagalore.

.... , : Annual Reports on Sericulture, Department of Sericulture, Kolar, 1985-86 to 1993-94.

Indian Institute of Economics : Techno Economic Survey of The Potentialities for the Development of Sericulture Industry in Andhra Pradesh, 1972. (VII Published).

International Trade : Silk Review, 1990-A Survey of Centre UNCTAD / GATT International Trends in Production and Trade, Geneva.

Institute for Social and Economic Change : Problems and Prospects of Sericulture, A Study in Some Village in Two Districts of T.N., J.Acharya and others, (Beneficiary Assessment Report), Bangalore, 1991.

National Institute of Rural Development : Block Plan in the District Frame, A Development Plan for Madakasira Block in Anantapur District, Andhra Pradesh, 1979.

National Commission on Agriculture : Report of the National Commission on Agriculture, Part III, Crop Production, Sericulture and Agriculture, New Delhi, 1976.

Prasad, B. : Studies on Reproductive Physiology of Silkworm Bomby x Mori (L) in relation to organic reserves of the mulberry leaves (unpublished thesis)

Reserve Bank of India : **Report on Financing the Crash Programme for the Development of Sericulture in Karnataka,** Bombay, 1974.

The Hindu : **Survey on Indian Agriculture,** Year Books 1989 to 1995.

Tariff Board, (Govt. of India) : **Report of the Indian Tariff Board Regarding Granting of Protection to Sericulture Industry,** Manager, Govt. of India Press, Calcutta, 1940.

Tariff Commission (Govt. of India) : **Report on the Continuance of Protection to the Sericulture Industry,** Bombay, 1974.

Tropical Development : Research Institute **The World Market for Silk,** Peter Green and Hangh, London, 1987.

ARTICLES

Abdul Majid : 'Dramatic Development in Andhra Sericulture', **Indian Silk,** Vol. XVII, No.6, October, 1978.

Arun Chandra, Guha : 'The Urgency of Cottage Industries,' **Khadi Gramodyog',** The Journal of Rural Economics, Vol.XXV, No.2, November, 1978.

Ashal, M.M. : 'Package and Practices for Mulberry Cultivation Under Temparate Conditions', **Indian Silk,** Vol.29, No.2, June, 1990.

Balasubramanian, V. : 'Sericulture as a High Employment Oriented Industry', **Indian Silk,** Vol.XXV, No.6, October, 1986.

Balwinder Singh : 'Impact of Local Points on Marketing of Farm Produce', **Kurukshetra,** 29(4), April, 1981.

Benchamin, K.V., & Jolly. M.S. : 'Employment and Income Generation in the Rural AreasThrough Sericulture', **Indian Silk,** Vol.XXV, June, 1987.

Benchamin, K.V. : 'Sericulture in Bangladesh', **Indian Silk**, Vol.32, No.6, 1993.

Bhatikar, A.P. : 'Sericulture and Rural Industrialisation (Part II)', **Indian Silk**, Vol.XXIV, No.3, July, 1985.

Bongale, V.D. : 'Mulberry Sericulture Zones and Agro-Climate in Karnataka', **Indian Silk**, Vol.30, No.5, September, 1991.

..........., : 'Mulberry Varieties in the Context of Indian Sericulture', Indian Silk, Vol.29, No.2, June, 1990.

Boraiah, G. : 'Establishment of Germplasm Bank in Mulberry and the Evaluation of Mulberry Varieties', **Lectures on Sericulture**, Suramya Publications, Bangalore, 1986.

..........., : 'Mulberry Cultivation', **Lectures on Sericulture**, Suramya Publications, Bangalore, 1986.

Central Sericultural Research & Training Institute, Mysore, : 'Mulberry Cultivation Under Rainfed Conditions A Challenge that should be met' **Indian Silk**, Vol - XXVI, No.81, December, 1987.

Chowkidar, V.V. & Pai, M.S. : 'Sericulture Development', **Economic Times**, May, 1976.

Datta, R.K. : 'Progress and Prospects of Sericul-ture in India Under the Central Sector', **Indian Silk**, Vol. XXV, No.9, January, 1987.

Dastagir, S.r. : 'Sericulture', **Land Bank Journal**, Vol.XVIII, June, 1980, Issue-VI.

Dayananda Reddy. R. : 'Farmers Training Programme An Effective Tool for the Transfer of Technology to the Field', **Indian Silk** Vol. XXVIII, No.8, December, 1989.

Dantwala, M.L. : 'Rural Employment - Facts and Issues', **Economics and Political Weekly**, Vol.XIV, No.2, June 23, 1979.

Devasurappa, L. : 'Silk Industry in Karnataka', **Reshme Krushi Journal**, January, 1980.

Deshmukh : 'Can Poverty be Removed from our Country' **Bharateeya Vikas**, Vol.1, No.2, October-December, 1980.

Editor : 'The Vesatality of Indian Silks', **Indian Silk**, Vol.VII, No.9, 1969.

..........., : 'Indo-Japanese Silk Accord Under World Bank Programme', **Indian Silk**, Vol.XIX, No.10, February, 1981.

..........., : 'Sericulture in India- Great Expectations and Great Challenges', **Indian Silk**, Vol.XXVI, No.9, January, 1988.

..........., : 'Chinese Leader Favours Silk Accord with India', **Indian Silk**, Vol.XIX, No.II, March, 1981.

..........., : 'Reports on Japan Sericulture', **Indian Silk**, Vol.XVI, No.4, August, 1977.

..........., : 'Sericulture comes of Age in India', **Indian Silk**, Vol.XXVII, No.1 May, 1989.

..........., : 'The Triumph of Sericulture in Andhra Pradesh **Indian silk**, Vol.XXX, No.9, January, 1992.

Gopalachar, A.R.S. : 'March of Sericulture Industry during Past Independence Era', **Indian Silk**, Vol.XI, No.4, Auguest, 1972.

George Fernandes, : 'India can Treble Her Silk Output', **Indian Silk**, Vol.XVII, No.7, November, 1978.

Hanumappa H.G. & Erappa, S. : 'Economic issues in Sericulture, Study of Karnataka' **Economic and Political Weekly,** August, 1985.

.... , : 'The Saga of Sericulture in the Princely State of Mysore', **Indian Silk**, Vol.XXVIII, No.4, August, 1989.

Hanumappa, H.G. : 'Rural Development Projects some Experiences with Karnataka Sericulture Project'- **Lectures on Sericulture**, Suramya Publications, Bangalore.

Hanumappa H.G. & Mangala : Issue in Sericulture Activities - Macro Perspectives'. **Lectures on Sericulture,** Suramya Publications, Bangalore, 1986.

Hanumappa, H.G. : 'Economics of Sericulture - Micro Perspectives', **Lectures on Sericul-ture**, Suramya Publications, Bangalore, 1986.

Iqubal, A., Badar : 'Performance and Prospects' **Khadi Gramodyog,** Vol.XXVII, No.11, August, 1981.

Jyoteeswar Patlik. : 'Silk Worm Rearing - A New Approach', **Yojana**, September, 1976.

Jayaswal, K.A., & Datta, R.K. : 'Some Popular Mulberry Silk Worm Races in India', **Indian Silk**, Vol.XXXI, No.9, January, 1993.

Kasivishvanathan. K. : 'The role of the state Khadi and Village Industries Board', **Indian Silk**, Vol.XXIX, No.4, August, 1990.

Krishna Swamy, S. : 'A New Strategy for Producing High Grade Raw Silk in Mysore', **Indian Silk**, Vol.X, No.1,1971.

.... , : 'Progress, Prospects and Problems of Sericulture in India', **Indian Silk**, Vol.XXV, No.12 & 1, April and May 14 1986.

.... , : 'Scope of Bivoltine Cocoon Crops in Mysore', **Indian Silk**, Vol.XI, No.9, January, 1973.

Kanyadi, N.M. : 'Sericulture in the Hill Areas of Karnataka, Encouraging Past Bright Future', **Indian Silk** Vol.XXVII, No.3, July, 1989.

Kamaraju Panthulu, N.: 'Development of Sericulture in Rayalaseema', **Khadi Gramodyog**, March, 1977.

Kanna, V.K. : 'Will China Resume the Silk Exports', **Indian Silk**, Vol.XXVII, No.7, November, 1988.

Lakshman. S. & Thiagarajan. V. : 'Growth of Indian Silk Export - Analytical Approach', **Agricultural Situation in India**, May, 1991.

Lele. D.V. : 'As Appraisal to Sericulture', **Khadi Gramodyog**, Vol.XXVII, December, 1979.

Manjeet, S. Jolly : 'Chawkie Rearing - Concept Organisation and Management' **Indian Silk**, Vol.XXV, January, 1987.

Mira Madan & Satyawati Sharma : 'Mulberry Sericulture in the Non-Traditional Areas', **Indian Silk**, Vol.XXX, No.5, October 1991.

Muneer Pasha, Md. : 'Swiss Aid to the Mulberry Sericulture Development in Andhra Pradesh and Tamil Nadu', **Indian Silk**, Vol.XXVII, No.3, July 1988.

.... , : 'The Road to Prosperity, Development of Bivoltine Sericulture in India', **Indian Silk**, Vol.XXVII, No.8, December, 1988.

Nataraja, N., & Mathew Thomas : 'Economics of Biroltine Sericulture in Karnataka', **Indian Silk**, Vol.XV, No.9, 1977.

Narasimhanna, M.N. : 'A Sound Seed Organisation for Sericulture Industry', **Lectures on Sericulture**, Suramya Publications, Bangalore, 1986.

..........., : 'Silk Worm Seed Production', **Lectures on Sericulture**, Suramya Publications, Bangalore, 1986.

Nagaraja Rao, H.A. : 'Facilities Extented to the Silk Reelers of Karnataka', **Indian Silk**, Vol.XXVII, No.9, January 1989.

Naidu, E.M., & Naidu, B.J. : 'Sericulture and Rural Development in Seventh Plan', **Southern Economist**', November, 1984.

Narayana, D.L. : 'Migration and Agricultural Develop-ment', **Khadi Gramodyogi**, July, 1966.

Patel, K.V. : 'Sericulture - An Instrument of Change - Some Gross Root Level Lessons', **Indian Silk**, Vol.XXXI, No.3, July 1992.

Patel, A.R. : 'Sericulture : A Labour - Intensive Industry, **Khadi Gramodyogi**, Vol.XXII, December, 1976.

Periswamy. K. : 'Problems and Prospects of Sericulture', **Indian Silk**, Vol.XVI, No.6, October 1977.

Rao, G.V.K. : Repeat World bank Sericulture Project for Speedy Progress', **Indian Silk**, Vol.XXIII, No.10 & 11, February 1985.

Rao, M.N. : Economics of Sericulture', **Khadi Gramodyogi**, Vol.XXII, June, 1977.

Ramakrishnan : 'Sericulture : An Evaluation of Impact', **The Economic Times**, November, 1987.

Rao, G.V.K., & Thamarakshi : 'Some Aspects of Growth in Indian Agriculture', **Economics and Political Weekly**, Vol.XIII, No.51 & 52, December 1978.

Rao, S.K., & Amul Snyal : 'On Promoting Employment through Labour intensity of techniques', **Economic and Political Weekly**, Vol.XIII, No.6 & 7, Annual Number, 1978.

Ranganatha Rao, K. : 'A Reportorial of an Inspiring Tour in Andhra Pradesh', **Indian Silk**, Vol.XVI, No.3, July 1977.

Royale, J.G. : 'Silk Culture in India', **Indian Silk**, Vol.XXV, No.2, June 1986.

Sanjay Sinha : 'Development Impact of Silk Production - A Wealth of Opportunities', **Economic and Political Weekly**, Vol.XXIV, No.3, January 1989.

Sankar, A. : 'Extension Agent - An Efficient Medium for the spread of Sericulture', **Indian Silk**, Vol.XXV, No.10, February 1987.

Sinha, J.N. : 'Rural Employment Planning - Dimensions and Constraints', **Economic and Political Weekly**, Vol.XIII, No.6 & 7, Annual Number 1978.

Shivananda, H.K. : 'Agricultural Markets and Cocoon Markets in Karnataka', **Indian Silk**, Vol.XXVI, No.5, September 1987.

Shoban Babu, E. : 'Economics of Silk Reeling and Twisting Units in Anantapur', **Indian Silk**, Vol.XXV, No.10, February 1987.

.........., : 'Sericulture Industry can Work Wonders for the Drought - Prone Area of Anantapur District', **Indian Silk**, Vol.XXVII, No.7, November 1988.

Susheelamma, B.N., & Benchamin, K.V. : 'Mulberry Tree Cultivation in Bangladesh', **Indian Silk**, Vol.34, No.2, June 1995.

Sonwalkar, T.N. : 'Factors Influencing Reeling Efficiency', **Indian Silk**, Vol.XVI, No.6, October 1977.

Thnang Dahnan, Linshi Xian & Lilong : 'Sericulture Production Strategies in the 21st Century', Linshi **Indian Silk**, Vol.33, No.8.

Thimmaiah, G. & Rao, V.M. : 'Problems and Prospects of Sericulture Development in Karnataka - A Field Review', **Sericulture for Rural Development.**

Tomy Philip : 'Why Silk is Precious', **Indian Silk**, Vol.XXVII, No.9, January 1989.

Vijayalakshmi et al. : 'Employment Generation through Sericulture in Andhra Pradesh', **Indian Silk**, Vol.XIX, No.10, February 1991.

Vijayalakshmi, G.S. : 'Sericulture the Queen of Rural Industry', **Yojana**, Vol.XXIII, No.21, 1979.

Veeraiah, K. : 'Sericulture in Karnataka, Yesterday, Today and Toworrow', **Indian Silk**, Vol.30, No.1, May 1991.

Vyas, V.Y., & Mathai, G. : 'Farm and Non-farm Employment Employment in Rural Areas - A Perspective for Planning', **Economic and Political Weekly**, Vol.XIII, No.6 & 7, 1978.

Index